£5.50
F

INTERNATIONAL SERIES OF MONOGRAPHS IN
PURE AND APPLIED BIOLOGY

Division: **ZOOLOGY**

GENERAL EDITOR: G. A. KERKUT

VOLUME 52

THE DETECTION OF FISH

OTHER TITLES IN THE ZOOLOGY DIVISION

THE DETECTION OF FISH

by

DAVID CUSHING

The Fisheries Laboratory, Lowestoft, Suffolk

PERGAMON PRESS

OXFORD · NEW YORK · TORONTO
SYDNEY · BRAUNSCHWEIG

Pergamon Press Ltd., Headington Hill Hall, Oxford

Pergamon Press Inc., Maxwell House, Fairview Park, Elmsford, New York 10523

Pergamon of Canada Ltd., 207 Queen's Quay West, Toronto 1

Pergamon Press (Aust.) Pty. Ltd., 19a Boundary Street, Rushcutters Bay, N.S.W. 2011, Australia

Vieweg & Sohn GmbH, Burgplatz 1, Braunschweig

Library of Congress Cataloging in Publication Data

Cushing, D H
    The detection of fish.

    (International series of monographs in pure and applied biology. Division: zoology, v. 52)
      Bibliography: p.
      1. Echo sounding in fishing.   I. Title.
SH344.23.E3C87   1973   639′.2′028   73–3446
    ISBN 0–08–017123–0

Printed in Great Britain by A. Wheaton and Co., Exeter

CONTENTS

PREFACE

MY INTEREST in echo sounding started aboard the R.V. Sir Lancelot in May 1948 off the coast of Northumberland and it has been maintained on other ships in various parts of the world including R.V. Ernest Holt, R.V. Clione, R.V. Platessa, R.V. Cirolana, R.R.S. Discovery, H.M.S. Hecla, R.V. Townsend Cromwell, M.T. Kirkella, and R.V./G.O. Sars. Much has been contributed by engineers employed by echo-sounder manufacturers and by scientists particularly in Norway, the United States and the United Kingdom. However, the most significant developments have often been made by the skippers of fishing vessels in many countries. Indeed the first steps were taken during the early thirties by a drifter skipper, the late Ronnie Balls.

The nomenclature I have used follows that in the "Plymouth Marine Fauna" (Marine Biological Association of the United Kingdom, Plymouth; 3rd ed., 1957), "A List of Common and Scientific Names of the Better Known Fishes of the United States and Canada" (Amer. Fish. Soc. Washington D.C.; spec. Publ. No. 1, 2nd ed., 1960), and certain other specialized publications. I am grateful to Dr. H. A. Cole CMG, Director of the Fisheries Laboratory, Lowestoft, for his encouragement and for reading the text. My colleagues, Dr. F. R. Harden Jones and Mr. R. B. Mitson have read the whole text and Dr. B. S. McCarthy has read Chapter 3 and their criticisms are gratefully acknowledged.

ACKNOWLEDGMENTS

PERMISSION to reproduce the following figures is gratefully acknowledged: Maxwell International Microforms Corp (Figs. 79, 80); the Editor of "Deutsches Hydrographische Zeitschrift" (Fig. 40); the Editor of "Discovery Reports" (Figs. 11, 21); the Comptroller of Her Majesty's Stationery Office (Figs. 8, 12, 13, 17, 18, 27, 28, 30, 33, 34, 36, 37, 38, 41, 43, 51, 69, 84); the Editor of the International Hydrographic Bureau (Figs. 26, 29); the General Secretary of the International Council for the Exploration of the Sea (Figs. 1, 3, 4, 9, 10, 12, 15, 44, 50, 55, 56, 57, 58, 59, 60, 61, 62, 68, 70, 73, 74, 88, 89, 90, 91); the Editor of "the Journal of the Acoustic Society of America" (Figs. 45, 46); the Company of Biologists (Fig. 42); the Editor of "the Journal of the Fisheries Research Board of Canada" (Figs. 54, 63, 64); the Editor of "The Journal of the Marine Biological Association of the United Kingdom" (Fig. 8); the Editor of "The Journal of Marine Research" (Fig. 31); the Editor of "Marine Bioacoustics", Vol. II, and Pergamon Press (Figs. 86, 87); Marconi Marine Ltd. (Fig. 48); the Marine Research Committee of California (Fig. 23); the Editor of "Nature" (Figs. 25, 53); Kelvin Hughes Ltd. (Figs. 35, 83, 85); the Editor of "Norwegian Fisheries Publications" (Fisk Dir. Skrift; Fisken/og Havet; Arsberet. ved Komm Norges Fisk.) (Figs. 4b, 5, 6, 14, 49); the Publications Committee of the Zoological Society of London (Figs. 81, 82); the Editor of "the Journal of Sound and Vibration" (Fig. 47); the Editor of "Protokolle für Fischereitechnik" (Fig. 75); John Wiley and Sons Ltd. (Fig. 78); the Editor of "Revue des Travaux de l'Office scientifique et technique de la Pêche Maritime" (Fig. 2); the Editor of the National Oceanic and Atmospheric Administration of the United States Department of Commerce (Figs. 7a, 7b, 20, 22); the Secretary of the International Commission for North West Atlantic Fisheries (Figs. 65, 66 and 67); the Editor of "World Fishing", and Kelvin Hughes Ltd. (Figs. 76, 77); the Editor of "Zoologica", New York (Fig. 19).

The authors' permission to reproduce the following figures is gratefully acknowledged: Monsieur C. Nédelèc (Fig. 2); Dr. F. Hermann (Fig. 4a); Dr. G. Saetersdal (Figs. 5, 14); Dr. G. T. D. Henderson (Fig. 8); Mr. J. Jakobsson (Fig. 10); Sir Alister Hardy F.R.S. (Fig. 11); Mr. A. J. Lee (Fig. 12); Dr. J. Eggvin (Fig. 15); Dr. J. H. Steele (Figs. 17, 18); Dr. J. E. King (Fig. 20); Dr. E. H. Ahlstrom (Fig. 22); Dr. F. R. Harden Jones (Figs. 42, 43, 88); Dr. R. H. Love (Figs. 45, 46); Mr. G. C. Trout (Fig. 53); Mr. I. D. Richardson (Fig. 69); Dr. K. H. Postuma (Fig. 70); Dr. J. Schärfe (Fig. 71); Dr. T. Aoyama (Fig. 72); Mr. A. C. Burd (Fig. 75); Captain C. Drever and Mr. G. H. Ellis (Figs. 76, 77); Dr. J. B. Hersey (Figs. 78, 79, 80); Mr. R. I. Currie (Figs. 21, 81); Dr. B. P. Boden (Fig. 82); Dr. F. J. Hester (Figs. 86, 87); Mr. A. R. Margetts (Fig. 93); Professor M. Uda (Fig. 16).

INTRODUCTION

THE first echo sounders went into service in the twenties but the most important step was the development of the recording echo sounder (Wood *et al.*, 1935). The first permanent records of fish were taken with machines of this type. During World War II, echo sounders and sonars were used for military purposes and after it, fishermen exploited them with great success for finding fish. In the first place military sounders and sonars were adapted by manufacturers for the purposes of fishermen. Today the machines are highly developed, with very much more power, higher resolution in range and angle, and much improved forms of display. There are specialized machines for different purposes, for finding fish near the bottom in deep water, ground discrimination, scientific work, deep oceanic sounding, fishing sonar and high resolution echo sounding. The development has been carried out by echo sounder manufacturers and the physicists and engineers who work for them.

Nearly all fishing boats in developed countries (and many in the developing countries) carry echo sounders of one form or another. In the main, fishermen have used the records from echo sounders as signs of fish. A few skippers will not employ them while others make highly sophisticated use of them. Some of the most important developments in echo sounding have been made by fishermen who saw much more in the records than the manufacturers believed possible. Some purse-seine skippers can forecast their prospective catches from the sonar records. Some trawler skippers use the detailed distributions of patches of fish quite close to the bottom in deep water to greatly improve the efficiency of fishing (Drever and Ellis, 1969). But the ideal state of affairs under which fishermen can make a quantitative assessment of the density of fish before they shoot their gear has not yet been reached. Attempts have been made to correlate catch and signal but they have often failed because of a number of sources of variability, amongst which is the natural variation in the density of fish in numbers per unit area.

Fisheries biologists are interested in numbers of fish per unit volume or per unit area because it forms the basis of their analysis of populations, and the catch or catch per unit of effort of a fishing vessel. Because the catches are variable and depend upon the presence of fishermen and their concentration, fisheries biologists have long desired an independent method of estimating the numbers of fish in the sea. However, their requirements are severe. They need to count the fish and identify them, but much more important they need to size them, because many fish grow by much more than an order of magnitude during their adult lives. The problem of identification appears at first sight to be of overwhelming importance but can be solved by capture. Then the function of the fishing gear is merely to provide evidence of identification. An ideal independent method of estimating fish populations is to express numbers of size groups per unit volume or per unit area. Then the survey of exploited populations and the exploration of unknown ones might become possible. This would then be quite independent of the presence of fishing fleets.

There are three groups of people important to the study of fish detection: the physicists and engineers who manufacture echo sounders and study the physics of sound in the sea;

the fishermen who use the machines in particular and specialist ways; and the fisheries biologists who would like to use them to count and size fish. The outlooks of the three groups differ considerably so that an echo sounder for a fisheries biologist looks and is very different from that used by a fisherman. The manufacturers of echo sounders have various views of the machines needed according to the markets available to them. The three view points have considerable areas of common ground, but the differences are real and originate in the historical development of the subject.

Chapter 1 in this book describes the various forms of fishery, how they are based on the biology of the fish, and particularly how the circuit of migration is expressed in the regularity of fisheries in time and space. It has been convenient to describe some distributions as echo surveys in this introductory chapter so that the reasons for carrying them out are put plainly in terms of the biology of the fish. Such surveys are referred to more fully in later chapters from other points of view, for example, as a stage in the history of the development of abundance estimation. At the end of the chapter two examples are given of the results that can be obtained from trawling surveys. Fish cannot yet be identified as species with acoustic equipment, but such equipment might estimate total biomass in unit volume quite easily. For some purposes, estimates of biomass per unit volume are valuable, but they are in general less valuable than samples of numbers by sizes of identified species. Consequently acoustic surveys are backed up by samples caught for identification and many fisheries biologists would prefer numbers and weights per unit volume to biomass per unit volume. Chapter 1 is directed mainly to the physicists and engineers who make echo sounders and who know little of fish biology or of the nature of the fisheries. Fishermen will recognize much that is familiar to them but they might find the biological aspects illuminating. The material must be familiar to fisheries biologists, even if they have not seen it arranged in this way.

The quasi-continuous record produced by an echo sounder appears to be a vertical section of the sea. Although the true depth of the sea is properly recorded, that of fish in the mid-water is usually over-estimated. In a broader sense, the echo record is deceptively simple because the flat paper does not appear to represent the volume sampled by the machine. Chapter 2 describes the nature of the paper record, the identification of fish. Fishermen often believe that the beautiful pattern displayed on the paper records indicates different species of fish. Although differences are recorded in shoaling pattern, in sizes and depths of fish, none can be relied on as being diagnostic. On a known ground at a particular season, fishermen expect to find the fish which they will catch. The association between catch and trace is then convincing even if it must fail elsewhere. However, some forms of identification are beginning to become possible. In some respects the echo record contains information which is under-exploited. For example, single fish can be distinguished from shoals, shoal size can be roughly estimated, fish speeds can be measured, and the average distance between individual fish can be estimated. They are measurements which might be of some value to fisheries biologists.

Chapter 3 states the acoustics needed to understand the sonar equations and the methods used for the estimation of absolute abundance with an echo sounder, which are presented in Chapter 4. The two chapters are written for fisheries biologists whose basic discipline is population dynamics. The end product in the procedures described is the estimation of numbers per unit volume, irrespective of the method used to obtain it. A number of means are now available and each is adapted to the special circumstances to which it has been applied. The last chapter describes some of the highly developed equipment which is now available. This equipment is complex and costly and may not come into general use, but the information collected with it will be used to develop the independent methods of estimating numbers.

The detection of fish is a study with a short history. In the last twenty-five years or so, echo sounders have been used by fishermen and fisheries biologists in all the oceans of the world. For the major part of the period, estimates of relative abundance were made primarily as a method of search. Inevitably, a method of search is subservient to other scientific ends. The most important development was the discovery that the target strength (or the standardized expected signal) of fish increased approximately as their volume. The consequence that fish could be sized preceded in history the discovery that numbers of fish could be expressed in unit volume. This discovery awaited the development of machines which displayed individual fish upon the paper record. It is only in the last five or ten years that the method of estimating absolute abundance has been considered a real possibility. Although echo sounders have other uses to fishermen and fisheries biologists, the detection of fish is on the verge of becoming a method in the study of population dynamics.

CHAPTER 1

WHERE THE FISHERIES LIE

1. *Introduction*

SINCE World War II fishermen have used the recording echo sounder and sonar to find fish. Nearly all fishing vessels carry acoustic equipment of one sort or another and fisheries biologists use the same gear to estimate abundance either relatively or absolutely. Both fishermen and fisheries biologists depend upon a regular seasonal pattern in distribution, because the fish tend to be found in the same place at about the same time of year. The animals migrate in a circuit recently described by Harden Jones (1968) in such a way that they are retained within the current structures. In temperate waters, it appears that fish spawn on rather restricted grounds at fairly precisely defined seasons (Cushing, 1969). After spawning, the adult fish drift, migrate, or diffuse away towards the feeding grounds, where they spread until they gather again for a spawning migration: the consistent time of spawning at the same place each year in temperate waters provides the basis of regularity in the fisheries.

Because of the regularity in migration, fish are usually found by the fishermen in expected places. Because the scale of search is so large, the search itself by capture or by detection plays a very important part. It will be shown that the art of fish detection has developed into the science of the acoustic estimation of abundance because the whole region in which a migratory circuit develops throughout the year can now be surveyed very rapidly. The early forms of detection depended upon capture only, and much of our present knowledge of the location of fisheries is based upon the collation of such information with that from echo surveys. As a background to the study of fish detection and as an example of the mass of information collected by mere capture and by echo survey, an introductory examination of the location of fisheries is given in this first chapter.

All trawlable areas in the North Sea are said to be "scraped over once a year" yet the fish are not scattered evenly but are found in patches, gathered for feeding or spawning. For example, plaice spawn in the early spring over high sandy ridges in the middle of the Southern Bight of the North Sea where they cannot be easily caught (Simpson, 1959). After spawning, the male fish move away north, close to Smith's Knoll Bank off the East Anglian coast and, near this position, they used to be caught by small trawlers working from Lowestoft. Another example is given by the arctic cod which migrate southwards in winter from the Barents Sea towards the Lofoten Islands in northern Norway. The fish move round the southern tip of the Lofoten Islands (Saetersdal and Hylen, 1959) and into the Vestfjord to spawn in February and March in a mid-water layer, and in this restricted area there is a heavy fishery on the spawning cod (Rollefsen, 1955). So the fish gather from all over the Barents Sea to spawn in an area about twenty miles long and such shoals have yielded a fishery for centuries.

Fish also gather on patches of their food and then the shoals are not as dense as the spawning ones. For example, during its oceanic life the Pacific salmon is caught by Japanese driftermen south of the Aleutian Islands and drift nets of 80 km in length are used. The aggregations of salmon are presumably small ones, which are spread out over the immense length of net. In contrast, at the mouth of the Fraser River, in British Columbia, Canada, gill nets for salmon are a few hundred meters in length. The difference in length of the nets between the oceanic fishery and that in the Fraser river is the difference in density between feeding and spawning shoals.

Fisheries biology is concerned with the study of fish stocks and it uses information in great detail. Historically, the first stage of research, which persists in the most sophisticated forms of study, is the rationalization and documentation of fishermen's knowledge. This has taken many forms, of which perhaps the most spectacular is the chart of the positions of catches of sperm whales (*Physeter catodon* Linnaeus) from 1729 to 1919 published by Townsend (1935) and reproduced here in Fig. 19: the original data were taken from the log books of catchers working out of Nantucket on the eastern seaboard of the United States. Incidentally, this chart is one of the best representations of the upwelling areas throughout the world: not only are the well known regions like the Peru current and the Benguela current clearly shown, but so are the equatorial complexes. Only comparatively recently has the form of upwelling along the equatorial region been understood and in the last decade a tuna fishery has developed there.

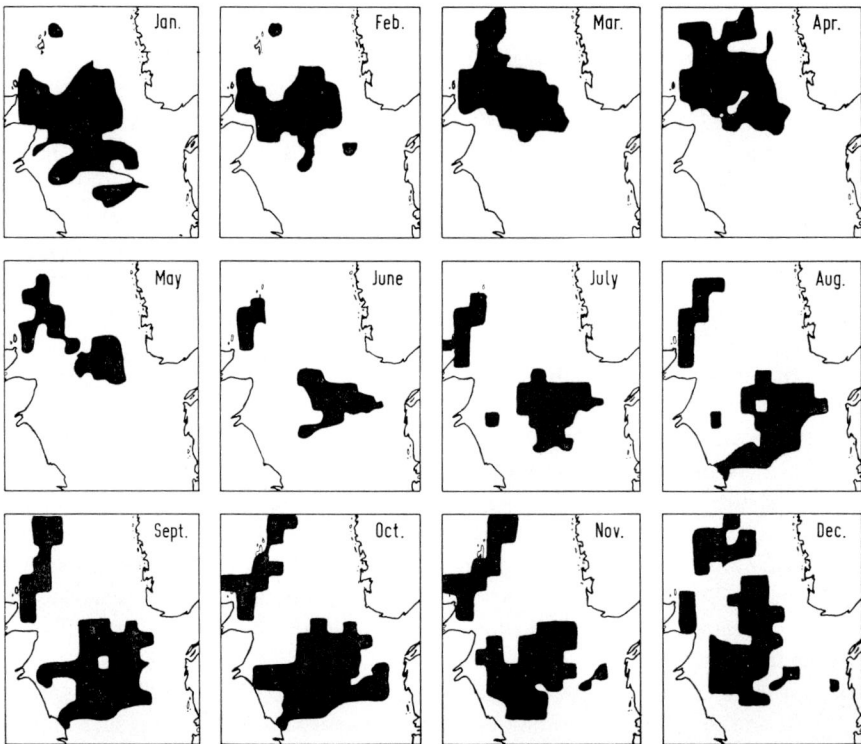

FIG. 1. Distribution of stock densities of haddock by statistical squares and by months in the North Sea
(Thompson, 1929).

A later stage of scientific development is the deliberate collection of statistics of stock density by small statistical squares and by short intervals in time. Provided that ships visit all areas in all months, the distributions of stocks can be charted adequately at different seasons. In the North Sea the fishery for demersal fish is intense but rarely concentrated; there is nearly always a trawler in sight, but rarely does one find the dense "cities" of trawlers common in Arctic seas. So it is possible to use charts of stock density in the North Sea to show where the fish are concentrated at different seasons. For example, Fig. 1 shows the distribution of the haddock stock densities (*Melanogrammus aeglefinus* Linnaeus) at different seasons of the year in the North Sea, because all the North Sea is fully sampled in all months. The changes were regular and rather slow; in the second half of the year the fish segregated into a northern and a southern group, which merged in the north in the following spring, the spawning season. It should be pointed out that Fig. 1 is based on data published in the thirties and that since then distributions in the fifties and sixties differed considerably.

The remainder of this chapter describes the location of some spawning fisheries and a few feeding fisheries, some at oceanic boundaries and a number in upwelling areas. The sources of regularity are revealed together with the scales of distance across which fish must be detected. But the acoustic equipment used today is complex and subtle and needs considerable insight to design it. Thus detection is not only the business of fishermen and fisheries biologists, but also that of engineers and physicists who make the equipment. For them this introductory chapter will describe the nature of the fisheries and some of the problems of fisheries biology. At the same time it will show how far echo sounders and sonar have been used to unravel these problems.

2. Spawning fisheries

(a) *The herring fisheries* (Clupea harengus *Linnaeus*) *in the southern North Sea*

There are a number of herring spawning grounds in the North Sea. Those of the Downs stock (Cushing and Bridger, 1966) in the Straits of Dover and Eastern Channel support fisheries in November and December. Figure 2 (after Ancellin and Nédelèc, 1959) shows the known spawning grounds; Galloper, Sandettié, Cap Gris Nez, Vergoyer, Creux St Nicolas and Ailly. The herring lay their sticky eggs on gravel banks of very limited area. The positions of these grounds do not vary from year to year; the evidence for this is, first, that spawning fish were always caught there, and, secondly, that in recent years, when the Downs stock has been much reduced, echo traces and patches of the youngest stages of larvae are rarely found at other positions. Indeed, since 1950, when herring trawling started in this area, the trawl fisheries have only occurred at the positions shown in Fig. 2. The area of a spawning ground is really small. That close to the Sandettié L.V. is about 3600 m along the tidal streams by 300–400 m across them (Bolster and Bridger, 1957); the evidence for the long term persistence of the grounds at the same positions is really in the location of the trawling ground, to which the trawlers return year after year, together with the fact that the aggregations of trawlers break up when the fish are spent (Ancellin and Nédelèc, 1959). In 1964 and 1965, new spawning grounds were found on the Bullock Bank 12 miles south of Beachy Head (Maucorps, 1966), but the herring also spawned in this area in 1951. These facts may provide evidence of the shift of spawning grounds or the discovery of minor grounds may only be an effect of the decline of the stock. It is interesting that an early

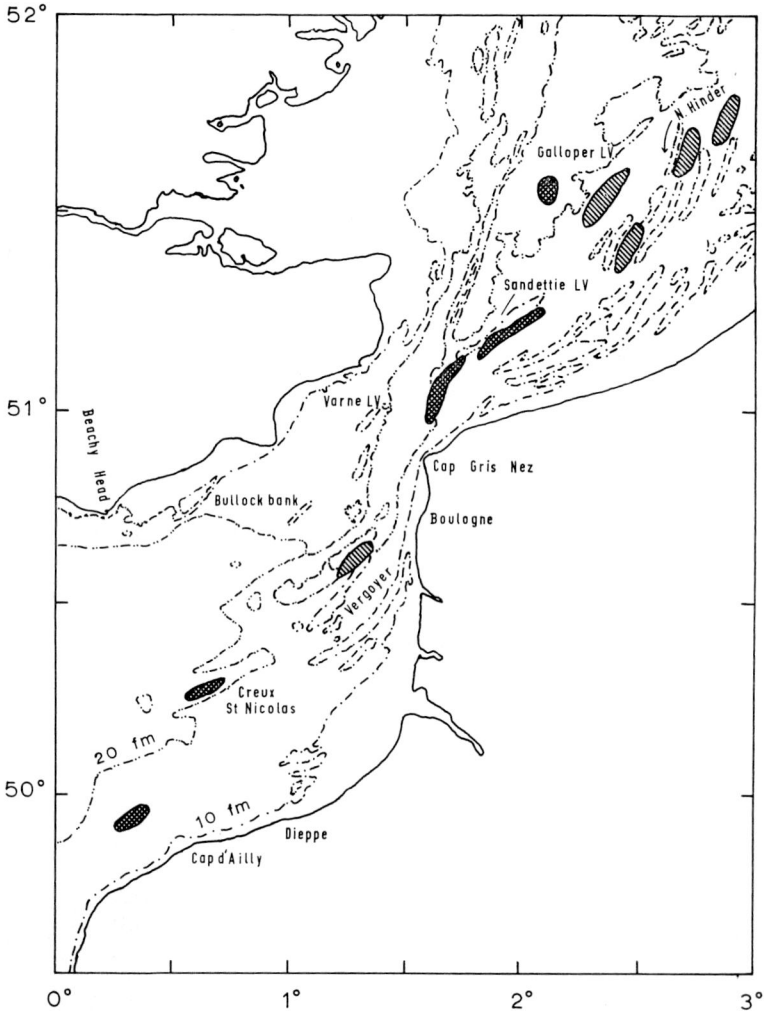

FIG. 2. The spawning grounds of the Downs stock of herring in the southern North Sea and in the eastern English Channel. The major grounds are Galloper, Sandettié, Cap Gris Nez, Vergoyer, Creux St. Nicolas, and off Cap d'Ailly; lesser grounds are found on the Bullock Bank, the Varne and on the Hinder Bank (after Ancellin and Nédelèc, 1959).

sonar survey (Renou and Tchernia, 1947) showed patches of signals in some of the areas charted in Fig. 2.

Herring approach their spawning grounds in the southern North Sea from the north (Burd and Cushing, 1962): they feed in summer in the northern North Sea between Fair Isle and Flamborough, mixing with other herring stocks that spawn off the Scottish coast in August and September and on the Dogger Bank in September and October. In September, fish of the Downs stock cross the Dogger area and appear off Smith's Knoll near Lowestoft in early October. Between the Knoll and the Brown Ridges off the Dutch coast they were caught by driftermen from England, Scotland, Holland, Germany and Poland. The fishermen shot a fleet of 70 to 90 nets, each about 14 m deep and about 35 m long, the total making a curtain about a mile and a half long. They were shot in the surface water just before dusk and were hauled in the early hours of the morning. The younger fish (of three and four years old) appear in October and the older fish (or six to nine years old) appear in November (Hodgson, 1957); the drift net fishery used to take place at the seaward ends of the Norfolk Banks, in the "swatchways" or channels between them, on the Brown Ridges, and above all, at Smith's Knoll. The English and Scottish fleets tended to work between the Smith's Knoll L.V. and the Indefatigable Bank (Erdmann, 1937) and the Dutchmen seawards of the British towards the Brown Ridges (Zijlstra, 1957). By the end of October and the beginning of November, herring were found in the southern part of the Southern Bight of the North Sea near the Gabbard L.V. and Galloper L.V. on the west side and on the Schouwen Ground on the east side. In November and December the fish were found on all the spawning grounds marked in Fig. 2. In the early months of the following year the spent fish drifted northwards from the spawning grounds.

Over the Flemish Banks, the pre-spawning aggregations were exploited by small pair trawlers from Belgium, Holland and France, using mid-water trawls. On the spawning grounds themselves, dense concentrations of large bottom trawlers were found in November and December. In the early fifties, the fishing season for the large vessels lasted as long as six weeks or two months, but as the stock became reduced, the season for the large vessels shortened considerably (Lundbeck, 1953, 1954, 1955) until in the early sixties the large trawlers stopped fishing there. The fishing at Sandettié and Cap Gris Nez ended as those on the Vergoyer and at Ailly started which suggested that although some fish spawned at the first two places, others passed through. This conclusion is borne out by a tagging experiment made by Ancellin and Nédelèc (1959). They also showed that fish tagged on the Vergoyer Bank and off Cap d'Ailly dispersed towards the French coast, where they were again exploited as spent fish by the pair trawlers. This fishery moved back through the Straits of Dover as if the spent fish were migrating back into the North Sea. Indeed in 1958 the pair fishermen followed the spent herring as far north as Texel Island off northern Holland which they reached in March (J. J. Zijlstra, personal communication).

In the twenties, there used to be a spring fishery for spent herring by drifters based on Lowestoft. It started in February off the Dutch coast, moved to the Brown Ridges and the Indefatigable Bank in April (Savage, 1930, 1931). Such a movement received a little confirmation from the recovery of two tags in the Silver Pit (just south of the Dogger Bank) in March from fish which were tagged near Sandettié a few months earlier (Bolster, 1955); in 1956, trawlers fishing the Silver Pits in March and April caught fish of the Downs stock on their way north (Schubert, 1958). The pattern of the fisheries for spent herring suggests a northerly movement because of the movement from the spawning grounds along the continental coast as far as Texel Island. Then the old Lowestoft spring fishery moved north from the Brown Ridges to the Silver Pits where in later years the fish of the Downs stock were caught by trawlers.

The herring were fished in the spawning, pre-spawning and the spent aggregations and the remarkable point is that they should have been found in the same positions from year to year. Those shown in Fig. 2 were found every season from 1950–1962 and to the north, in the former area of drift net fishing in the Southern Bight, there were similar areas shown by catches and echo patches which were more extensive, less well defined, and more variable in position from year to year (Valdez and Cushing, 1962), but nevertheless, were found in the same region in each year. It is this regularity that sustains a fishery.

(b) *Norwegian spring herring fishery*

Until the late fifties the Norwegian spring herring spawned on the coast of Norway in about 40 fm (Runnstrøm, 1936) particularly off Haugesund and the island of Utsira. Figure 3 shows a summary of the seasonal migrations of the Norwegian herring (Devold, 1963).

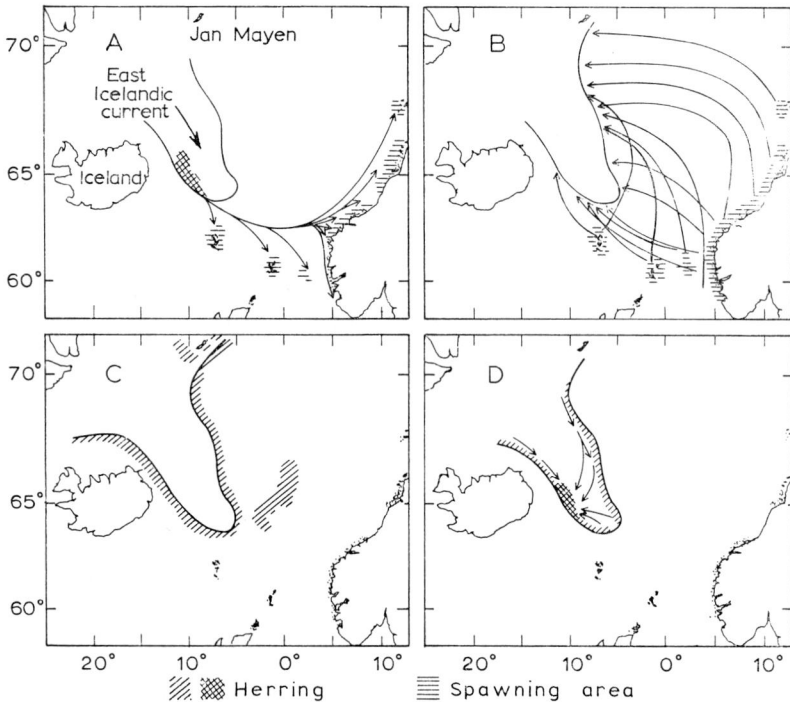

FIG. 3. The migration of the Norwegian herring: (A) the movement of fish in the East Icelandic current shown as a tongue towards the Faroe Is., Shetland Is. and the coast of Norway in late autumn and winter; (B) the movement of spent fish across the Norwegian Sea towards the feeding area between Iceland and Jan Mayen; (C) herring in the feeding area; (D) the movement of herring southward to the area east of Iceland before the main spawning migration (Devold, 1963).

The spawning area lies on the coast of Norway, and outlying stocklets spawned off the Faroe Islands, the Shetland Islands and on the Viking Bank. Spawning takes place in February and March. After spawning, the fish moved off into the Norwegian Sea and found their way to the polar front between Jan Mayen and Iceland. Here they fed and in late summer they moved back towards the Faroe Islands and the Norwegian Sea in the East Icelandic current. Figure 4 gives the detailed evidence of this movement in mid-winter from the Faroe Islands to the Norwegian coast. Figure 4a shows a large patch of herring traces observed by sonar in the confluence of two oceanic currents north of the Faroe Islands (Hermann, 1952). If the herring drifted with the currents they would be expected to move eastwards and the north-eastwards with warm Atlantic water over the Wyville Thomson ridge between the Faroe Islands and the Shetland Islands. Figure 4b shows the patch of fish traces during the winter of 1950–1951 at the mixing area of the East Icelandic current and the Atlantic water. They swim under the warm water and reach the Norwegian coast north of Bergen in January (Devold, 1966) and make the crossing from the Faroe

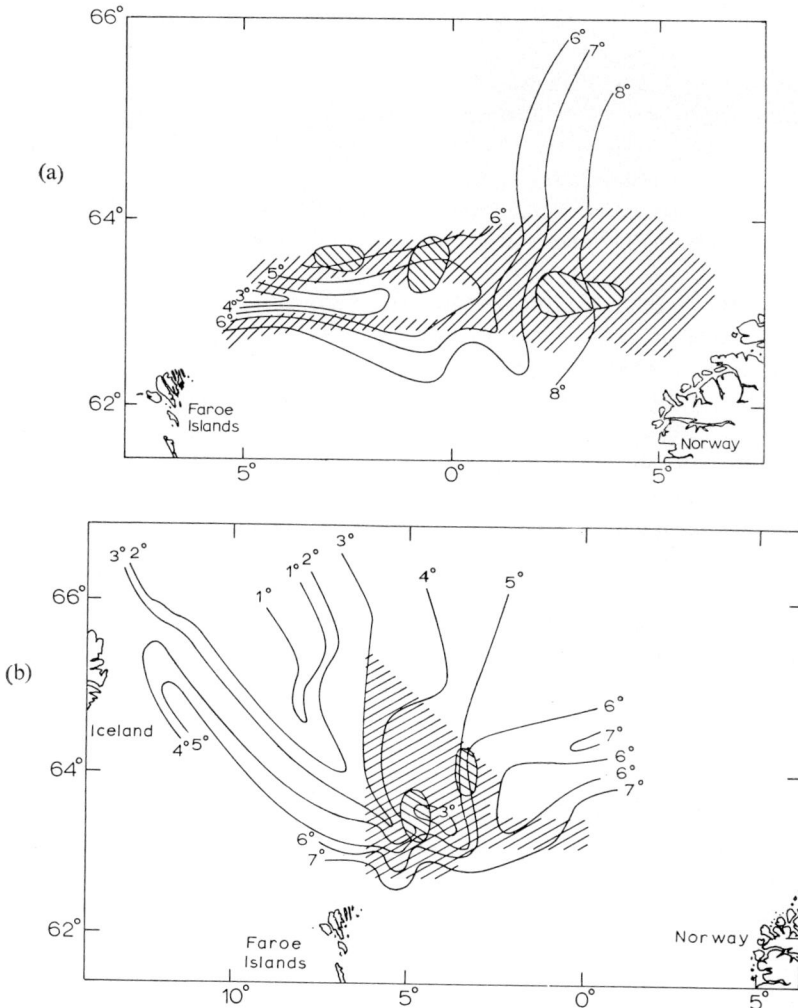

FIG. 4. a. Distribution of echo traces of herring north and north east of the Faroe Is. in December 1951 and January 1952 (Hermann, 1952).
b. Distribution of echo traces of herring north east and east of the Faroe Is. in December 1950 and January 1951; the current boundary is shown by the temperature distributions and the fish appear to have partly crossed it (Devold, 1954).

Islands to Norway in seventy days. This migration under the Atlantic stream was discovered by a sonar survey.

The distributions of large herring and spring herring were charted on the Norwegian coast during the thirties (Runnstrøm, 1936). The spring herring are those on their spawning grounds and the large herring are the pre-spawning aggregations to the north. It is as if the fish on their way from Iceland struck the coast of Norway well to the north of the spawning grounds and then moved south to spawn. The positions of spawning grounds off Haugesund were revealed in detail by grab surveys (Runnstrøm, 1941). Off Haugesund lies the island of Utsira; around it and between it and the mainland hundreds of grab samples were taken, about a third of which were positive. Runnstrøm said that the fish moved south in the Norwegian deep water and that they moved inshore in the deep channels cut in the landward wall of the Norwegian Rinne to find particular spawning grounds. He produced evidence to show that the time sequence of the appearance of eggs on the spawning grounds is the same from year to year, i.e. that some are February grounds and that others are March grounds. This evidence suggests that Norwegian herring, like North Sea herring, spawn regularly on grounds of limited extent. After spawning, the large spring herring move off northwards and Russian work (Marty and Wilson, 1960; Fedorov, 1960) has suggested that they appear next in their feeding grounds in the center of the Norwegian Sea, as suggested in Fig. 3.

The fishery for the large and spring herring before World War II was mainly carried out by drifters in the north of the area and by sunk nets (or bottom gill nets) in the area of the spawning grounds. Since then, this fishery has been replaced by a purse-seine fishery working over the whole area which has shifted north from the Haugesund region to the Lofoten Islands. It has taken place from year to year on the spawning migration. It is one of the triumphs of Norwegian fisheries research to have shown by sonar survey, not only to scientists but to fishermen each year, when and where the shoals of large herring struck the Norwegian coast (Fig. 4a and b). Since 1958, the spawning fishery has moved north as the 1950 year class grew older. The later year classes, 1959 and 1960, appear to have spawned outside the Lofoten Islands where they were located by sonar survey (Devold, 1965). Catches off the Norwegian coast have alternated with those off the Bohuslån coast of Sweden in a periodic manner since the Middle Ages and each period lasts about 50–70 years. Aasen (1962) thinks that the shift in time towards later spawning and the northward shift in space indicates the end of the recent herring period, but recently, it has been suggested that the northern stocklet represents a separate stock component.

There is no doubt that the Norwegian and Swedish catches have alternated for many centuries. But there is no evidence that the catches are made from the same stock; indeed Höglund 1960 has produced evidence from fossil vertebrae that the fish caught off Sweden were really North Sea spawners. Runnstrøm's evidence of local and constant spawning grounds suggests that the Norwegian herring catches were derived from an array of stocklets. Then the observed shift in spawning season and of spawning grounds would really be a shift from stocklet to stocklet as an effect of changes in abundance between them. This shift might well be an effect of climatic change because Beverton and Lee (1965) have correlated the alternation of periods with changes in the ice cover off Iceland, which indicates climatic change.

There is sufficient regularity in the spawning fisheries of the Norwegian herring to allow the fishermen to find them year after year. During the 1920's and 1930's the fisheries were regular enough for Runnstrøm to establish the main biological features of the stock. Then, during the 1950's the form of spawning migrations was established by sonar survey and indeed the sonar survey demonstrated the northward shift of that migration.

(c) *The Lofoten cod fishery* (Gadus morhua *Linnaeus*)

Herring spawn on the bottom and their eggs stick to gravel patches and the ripening fish migrate to fixed spawning grounds of limited extent. Cod spawn in the mid-water and as their eggs are pelagic, they drift away from the spawning ground in the currents (Sars, 1864/9). Hence the precise location of a major spawning ground of the Arcto-Norwegian cod stock in the Vestfjord (in northern Norway) in rather a small area is a surprise. It is true, however, that some fish tagged in the Barents Sea have been recaptured at some consider-

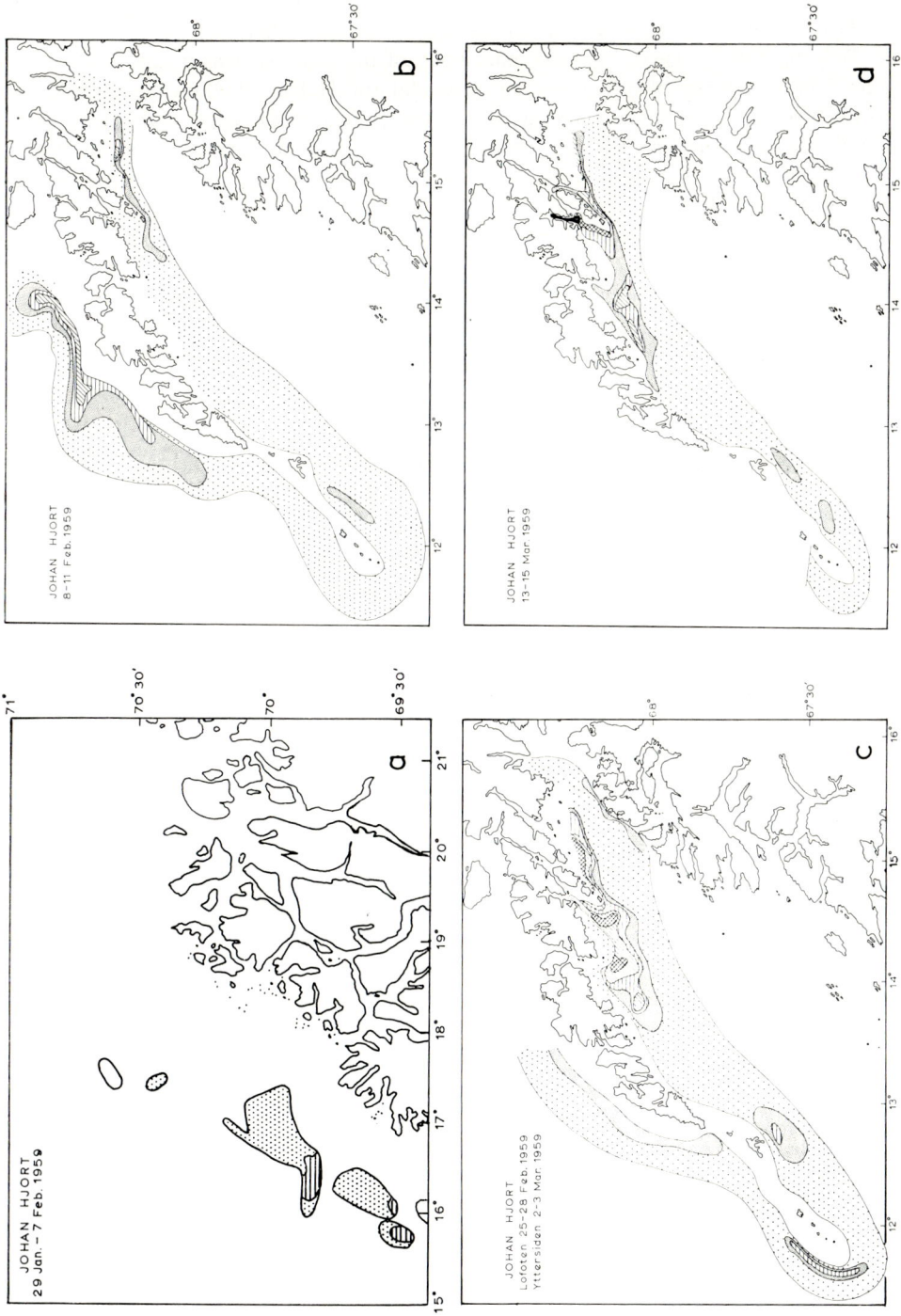

FIG. 5. The approach of cod to the Vestfjord in northern Norway in January, February and March as shown by echo survey (Saetersdal and Hylen, 1959).

northwards along the west coast of Hokkaido and even on to the north east coast of that island by 1810. The fishery, which has declined since 1904, is now restricted to the north east coast of Hokkaido. The long term of records show that the spawning fishery persisted for hundreds of year along the western coast of Hokkaido. There was a persistent regularity, lasting for centuries until heavy fishing supervened at the turn of the last century.

(e) *The Pacific albacore fishery*

Albacore, *Thunnus alalunga* Bonnaterre, are caught in two main regions of the North Pacific anti-cyclone, the North Equatorial current and the North Pacific current, or the Kuroshio and the Kuroshio extension. Suda (1963) believes that one stock lives in the North Pacific gyre and another in the South Pacific one (Fig. 7a.) Fish tagged off California have crossed the ocean in less than a year (Otsu, 1960). There is some evidence in size and in morphological characters of difference between stocks in the northern and southern gyres (Suda *et al.*, 1963), but there is little evidence of stock differentiation within the northern anti-cyclone.

The spawning fish, found in the North Equatorial current, are larger than 90 cm. In the Kuroshio and the Kuroshio extension, the fish are mainly immature. The large spawning fish are caught in the North Equatorial current by Japanese long liners in June and July in a true spawning season that is probably much more extensive. In the South Equatorial current, spawning fish are caught between October and March, peaking in December. The fishery shifts westerly with time as if the fish were drifting westwards with the current. It is very interesting to find that the spawning seasons of the two stocks are six months out of phase with each other. There is little seasonal difference at the equator, but there is considerable difference in season between the two anti-cyclones, treated as a whole.

The average age in long line fisheries at the equator is four, ranging from three to six. The larvae of albacore when under 15 mm cannot be properly identified (Yabe *et al.*, 1963) but larvae which may well be albacore have been found in numbers in the North Equatorial current. Indeed those that have been properly identified were found in this current (Yoshida and Otsu, 1963) and albacore juveniles are found in the same current. Skipjack and Yellowfin larvae are found in the divergences and convergences associated with the South Equatorial current and the Counter-current.

It is probable that the migration circuit of the albacore is based on the oceanic anti-cyclones in the Pacific and the adult fish of northern and southern stocks probably spawn in summer in the Equatorial currents, north and south. In Fig. 7b, the distribution of catches in February is shown in relation to the oceanic current system. In point of scale alone, the distribution of the albacore is oceanwide. Each current is hundreds of miles wide and thousands of miles long, but this is a relatively small area in the largest ocean. An albacore is not a very large animal, not more than 120 cm in length. The spawning ground is a very long strip and the spawning season may be very long, because on such a scale, the regularity of spawning must be difficult to maintain. Further, although it has been recently shown that in temperate waters the peak spawning dates are rather precisely determined, in the sub-tropical anti-cyclones they are loosely determined (Cushing, 1969). But the Japanese fishermen have established enough regularity in the system to catch the large fish in the Equatorial currents and the small fish in the Kuroshio extension and its South Pacific analog.

(f) *Summary*

Five spawning fisheries have been described, the East Anglian herring fishery, the Norwegian spring herring fishery, the Lofoten cod fishery, the Hokkaido winter herring fishery, and the Pacific albacore fishery. Each fishery has a regular season: the peak dates of two of them do not vary very much, by perhaps a week or ten days, and this may be a general phenomenon in temperate seas. For the Pacific albacore fishery, the regularity is merely the appearance of large fish in the North Equatorial current during the northern summer.

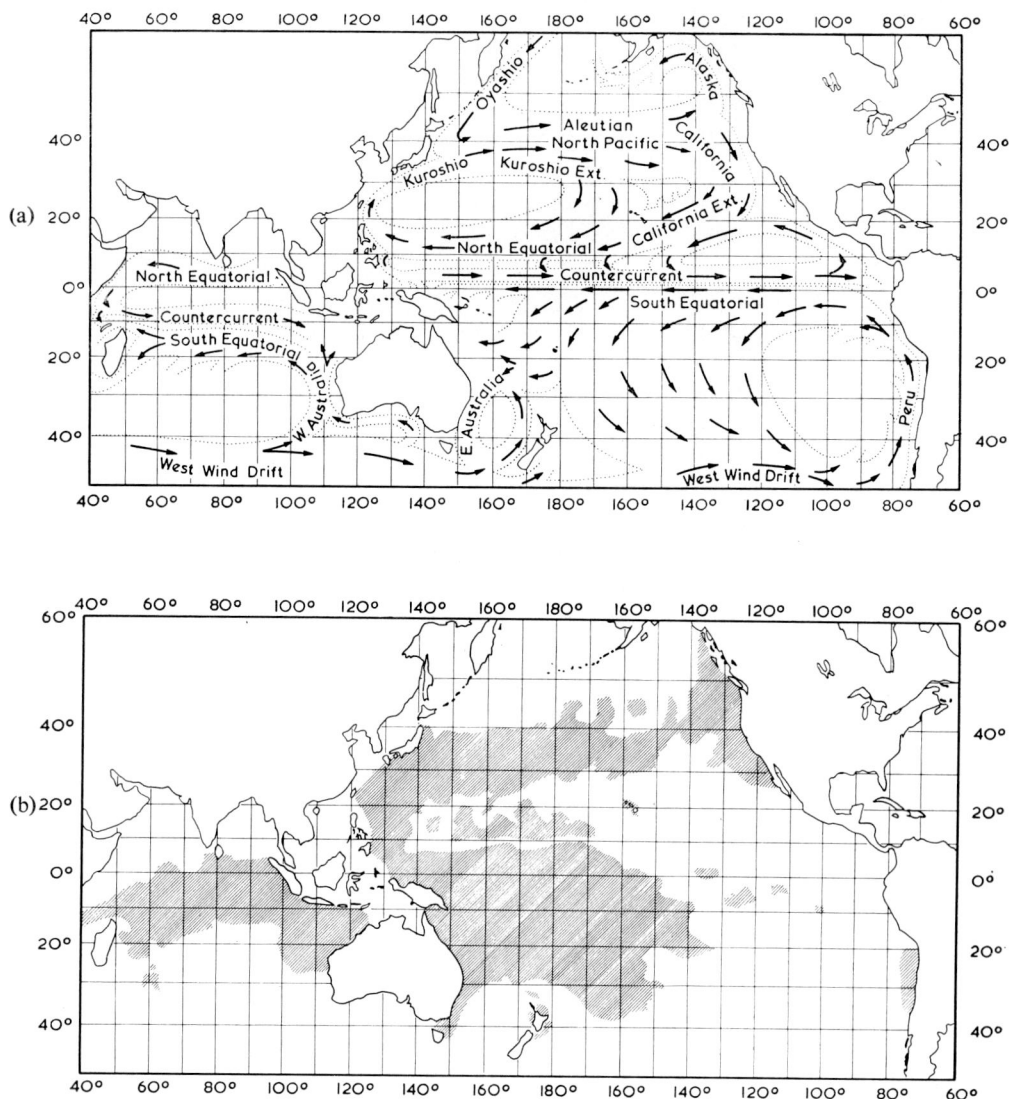

FIG. 7. a, The surface current system in the Pacific Ocean (Brock, 1959).
b, Catches of albacore in the Pacific and Indian Oceans (Brock, 1959).

The herring spawn on closely defined patches of seabed in water as deep as 40 fm (*C. harengus*) or as shallow as the gravels and weeds (*C. pallasi*) between the breakers and so there might be good reason for their annual return to the same place to spawn. There is an obvious analogy here with the Pacific salmon which returns after years to its parent stream.

The cod spawn in a mid-water layer in the Vestfjord. To do so they migrate from Spitz-bergen, Bear Island, Hope Island, Vardo, and Skolpen. How they do this is unknown, but in the process they reach the Vestfjord where the majority spawn in a mid-water layer, the position of which is predictable from year to year. The albacore, however, lives in mid-water all his life, spawns in mid-water, and travels in the North Equatorial current during the spawning season. But the regularity of spawning season is relaxed and there is no precise spawning ground other than the long current on the Line which of course is a small pro-portion of the North Pacific anti-cyclone.

To understand the location of spawning fisheries and to probe their regularity, we need two forms of information. First, the migration circuit of the whole stock must be known in some considerable detail. Secondly, we have to know how the fish return so regularly to spawn under such an extensive variety of conditions. In some fisheries, particularly those described here, the migration circuits are known and it is understood that they are contained within the current systems. For example, the Norwegian herring moves round the Norwegian Sea roughly in the direction of water movement. This is sufficient to describe the habitat, and the circuit of its migrations. But the mechanisms remain undiscovered and the small differences in the spawning migrations from year to year are not yet understood.

In addition to the problems of migration, there are also those of maturation in the broadest sense. Not only are we concerned with the ripening of gonads and act of spawning, but how maturation is seasonally modulated in temperate water. The precise timing of spawn-ing seasons requires some accurate means of recording the passage of the seasons; birds do it by measuring day length and perhaps fish do the same. Equally fascinating is the relaxa-tion of the precise control of spawning season in the tropical and sub-tropical ocean; the clue to the difference is probably the continuous production cycle of the tropical ocean, as contrasted with the discontinuous one of temperature waters.

3. Feeding fisheries

Spawning grounds in temperate waters are often found in the same place each year. In the feeding fisheries the problem is different, based on the dynamics of aggregation and disengagement of shoals on to food patches. Consequently, the areas in which fish are to be found are much more extensive.

(a) North Sea herring

The first systematic attempt to describe a feeding fishery was that made by Hardy and his collaborators off North Shields and off Fraserburgh in the North Sea during the early thirties (Hardy et al., 1936). Herring were caught by drifters and some drifter skippers were given plankton indicators. They are little torpedo-shaped tubes which were towed for a brief period at full speed. The plankton was collected on silk discs set across the end of the tube. The drifter skippers were taught how to recognize Calanus finmarchicus Gunner, the preferred food of the herring.

At first, plankton indicators were shot randomly with respect to the distribution of Calanus and of herring. So shots of herring were recorded in poor Calanus areas as often as in rich Calanus ones. A typical set of results is shown in Fig. 8. Single shots of herring are shown as black histograms ranked from left to right in order of Calanus numbers which are also ranked, lowest left and highest right. The distribution of Calanus numbers is split

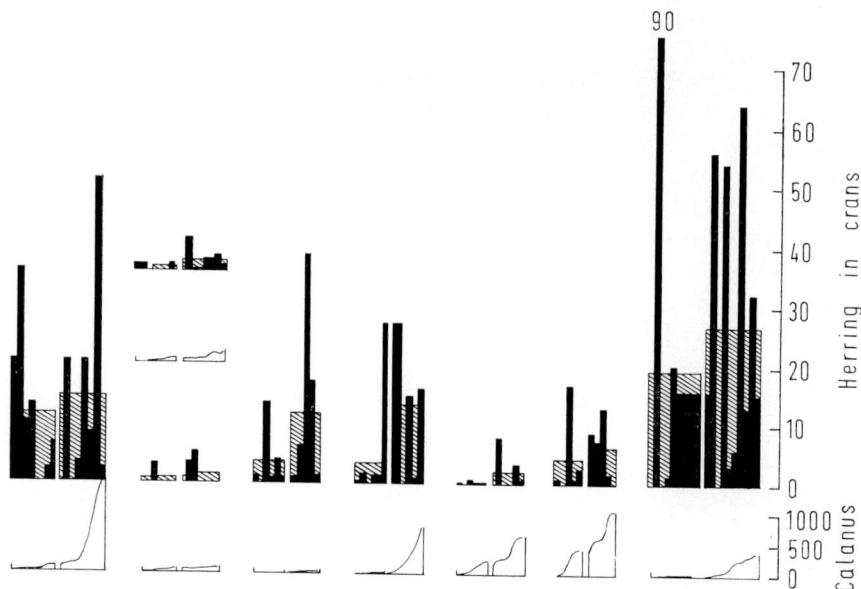

FIG. 8. Catches of herring in "rich" and "poor" *Calanus* are ranked in ascending order of numbers (full line); the mid-point splits the rank into "rich" and "poor". The black histograms give the catches made at each value of *Calanus*. The hatched histograms give the average catches in "rich" and "poor" *Calanus* (Hardy *et al.*, 1935).

at the mid-point into "rich" and "poor" *Calanus*. The average catch of herring shown hatched in the figure in "rich" *Calanus* was always higher than that in "poor" *Calanus*. In the published data (Hardy *et al.*, 1936), gains in catch do not always occur, losses sometimes prevailing. But a low average gain would sustain profits to a fleet. Yet the technique was not adopted by British fishermen and the reason can be seen readily in Fig. 8 because the variability of catches (the black histograms) is very high.

A large number of shots would have to be dragged over the side before the skipper and his crew would see the result so obvious in Fig. 8. The *Calanus*/herring relationship is transient, lasting perhaps three or six weeks in a given area and so the skipper cannot for himself establish the relations shown in Fig. 8 for the fleet.

But one skipper used the plankton indicator with success. The late Skipper Ronnie Balls has described (Balls, 1954) how he steamed all one summer day for a great distance in the northern North Sea shooting and hauling the plankton indicator and finding nothing but the stinking green of dense diatom patches. In the late afternoon he found a small dense patch of *Calanus* clean of diatoms, each animal kicking and "strong alive". He steamed beyond the patch and again found the stinking green and so returned to the patch of *Calanus*, shot his nets and caught 150 crans of herring, (nearly 30 tons). The herring were there because they had gathered on a *Calanus* patch and the *Calanus* were there because they had eaten the diatoms, had reproduced and had locally increased their numbers.

A feeding fishery in temperate waters has its own dynamics. During the spring algal outburst, *Calanus* and the other copepods feed and reproduce. As their numbers increase they reduce the numbers of algae by grazing. In time, the *Calanus* peak follows the diatom peak and in space, *Calanus* patches replace diatom patches. If herring slow down or stop to feed, a mass will gather onto the *Calanus* patches. This is sufficient to explain the results given in Fig. 8 and to explain the fact that herring are not caught in the diatom-rich areas, as shown by Skipper Ball's experience. So long as the density of *Calanus* is sufficient to slow down the feeding herring, the herring numbers will increase. The *Calanus* numbers do not increase very quickly naturally and so the herring tend to reduce them. Their numbers decline and the herring will leave the area because the fish disengage more quickly than they gather (Cushing, 1955). *Calanus* numbers and herring catches are positively related in the early stages of the dynamic process. As *Calanus* numbers fall, *Calanus* and herring tend to be inversely related. Thus, the best herring catches are made just before the fish swim away. Savage (1937) showed that *Calanus* was the commonest food of the herring in the North Sea in April, May and June, the period of ravenous feeding. Figure 9 shows the relationship

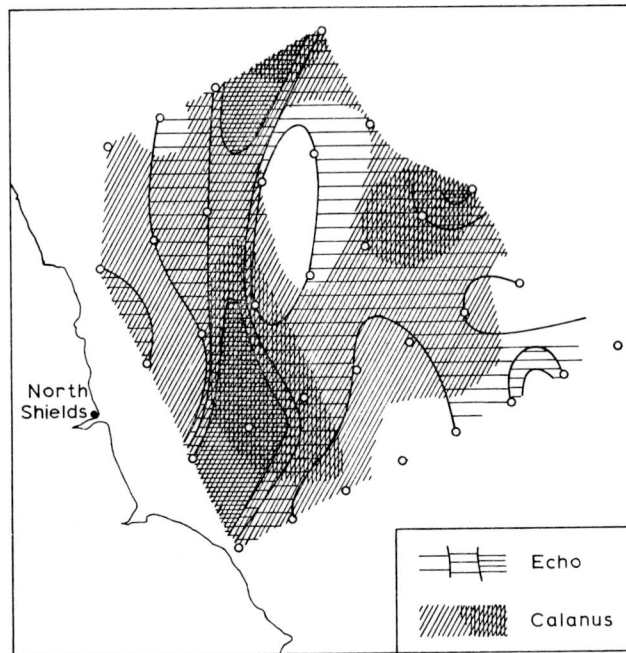

FIG. 9. Aggregation of the echo traces of pelagic fish (probably herring) on to patches of *Calanus* in May 1949 (Cushing, 1952).

between the density of *Calanus* off the north east coast of England in 1949 and the density of echo traces recorded at the same time (Cushing, 1952). It is likely that most of the traces were recorded from herring at this time. The process of aggregation of herring on to *Calanus* patches was roughly described using the correlation coefficients (See Table 1) between echo density and that of various copepods for five successive cruises off the north east coast of England in 1949. Each coefficient expresses the degree of aggregation shown pictorially in Fig. 9.

As a measure of aggregation the significance of the correlation coefficient is of limited value because a low coefficient is as meaningful as a high one; the significant correlations show that positive aggregation has occurred, but the low correlations showing no aggregation are equally useful.

TABLE 1. THE AGGREGATION OF PELAGIC FISH (AS DENSITY OF ECHO TRACES) ON DENSITY OF COPEPODS IN APRIL AND MAY, 1949 (Cushing, 1955)

				Cruise		
		VI	VII	VIII	IX	X
Copepod						
Evadne nordmanni Lovén		—	—	—	—	0.18
Temora longicornis (OF Muller)		0.54**	0.25	−0.02	−0.10	0.17
Pseudocalanus elongatus						
(Boeck)	male	0.08	0.04	0.04	−0.12	0.04
	female	0.16	−0.17	−0.13	−0.16	−0.25
	juveniles	−0.18	−0.26	0.15	−0.01	−0.05
Paracalanus parvus (Claus)	adults	−0.12	−0.06	0.15	−0.20	0.09
	juveniles	−0.12	−0.20	0.22	−0.05	0.08
	Copepodite stage					
Calanus finmarchicus	V	−0.32	−0.34	0.44**	0.25	0.13
	IV	−0.38	−0.15	0.20	0.30	0.70**
	III	0.29	0.22	0.45**	0.16	0.42
	II	0.20	−0.32	0.17	0.31	0.32
	I	0.14	−0.26	0.32	0.16	0.25
	(IV and V)			0.32	0.20	0.54*

Note: Significant difference from zero is noted as * ($P = 0.05$) and ** ($P = 0.01$).

In Cruise VI there is a positive aggregation on to *Temora longicornis* and in Cruises VII, and X on to the *Calanus* patches, stages III, IV, V, whereas in the earlier cruises there is no such relationship. It is possible that the fish first aggregated on to *Temora* and then on to *Calanus*, when the latter became more abundant. The cruises were about ten days apart and the aggregation on *Temora* lasted perhaps three weeks. That on the older *Calanus* copepodite stages may have lasted about a month. So during the period of ravenous feeding in May there is a short time when catches of herring might be indicated by "rich" *Calanus*.

The North Sea herring feeding fishery in May has been treated at some length to illustrate the transient and dynamic nature of the processes. Any feeding fishery depends on a number of factors. The food density must be much higher than the low limiting density at which the searching fish slow down. There must be enough fish elsewhere to build up the herring numbers in the *Calanus* patch. The rate of aggregation must not step up the mortality of the food too quickly or the fish will disengage. Hence, roughly, the more food, the greater the chance of a feeding concentration enduring.

(b) *The Icelandic herring feeding fishery*

The main feeding fishery for herring takes place off the north and east coast of Iceland in July and August. The fishery continues in the East Icelandic current until December when it is more a migratory fishery than a feeding one.

However, it starts as a very important feeding fishery, which until very recently landed about 0.5 M tons a year. The herring feed mainly on *Calanus finmarchicus* as in the North Sea. When the fish feed on euphausids, poor catches are expected and when they feed on fish larvae, the fishery fails (Jakobsson, 1963). Figure 10 shows the relationship between herring patches as recorded by sonar traces from the Icelandic research vessel Aegir and the zoo-plankton (mainly *Calanus*) sampled in the top 50 m of the sea. On the first cruise in early July there is no relationship between *Calanus* and the sonar traces. On the second and third cruises (the third cruise is illustrated in Fig. 10) nearly all the traces are amongst, or very close to, zooplankton patches. On the fourth cruise, the zooplankton has been reduced save for one small patch in the west and one in the north; on the last two cruises it is in these two limited areas that sonar traces are found. The density of shoals is determined by the density of the zooplankton patches and, as the *Calanus* is eaten down, the fish disperse and with them the fishery. The whole process of aggregation and dispersal to a south easterly migration lasts for just about two months.

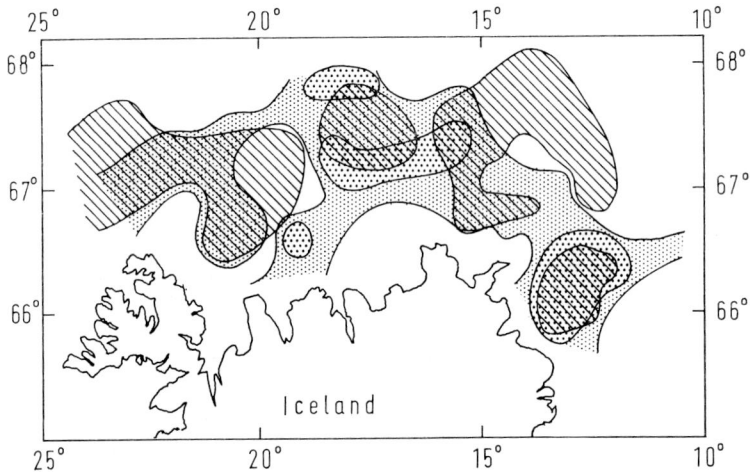

FIG. 10. Aggregation of sonar traces on to patches of *Calanus* in the feeding fishery north and north east of Iceland; the dotted areas are patches of zooplankton and the hatched areas are patches of sonar traces (Jakobsson, 1963).

The *Calanus* densities are much greater than those in the North Sea off North Shields and the herring stock supplying this fishery, the Norwegian herring, was a bigger one than those found in the North Sea. Hence the fishery continued for a longer period than the transient feeding fisheries in the North Sea, as expected from the rough description of pro-cesses given in the previous section. During the sixties, the Icelandic summer feeding fishery has been very intense. It was this heavy fishery that allowed the scientists to confirm decisively that the sonar traces were really records of herring. Much effort is directed to searching for herring by the Icelandic scientists. Jakobsson, who led the search, relied on sonar traces to find the fish, but used the *Calanus* to show how the fishery would proceed.

(c) The whale fisheries

Whale fisheries are feeding fisheries and the most spectacular were those formerly based on the high latitudes. The sei whale (*Balaenopter borealis* Lesson) feeds almost exclusively on *Calanus finmarchicus* ("red water"). Hjort and Ruud (1929) have shown that the number of sei whales caught rises and falls subsequent to the rise and fall in the quantities of *Calanus* found in the water, and the temporal sequence is shown well. The aggregation of whales builds up and then the *Calanus* numbers are reduced by whale grazing. Then the whales disengage and the transient process is completed. The spatial distribution off West Greenland showed that catches of blue (*Balaenopter musculus* L.), fin (*Balaenoptera physalis* L.), sei, humpback (*Megaptera novea-angliae* Borowski), and sperm whales were related to the numbers of euphausids in the top 50 m of the water (Hjort and Ruud, 1929). So the whales were caught on the bank where the euphausids had gathered.

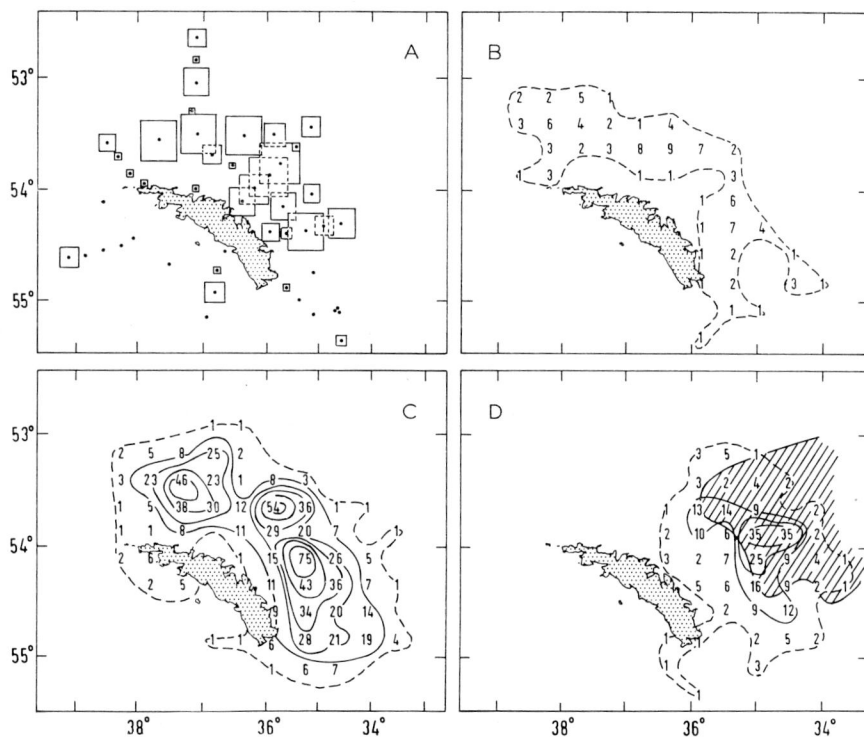

FIG. 11. Catches of whales and the distribution of *Euphausia superba*, north of South Georgia (December 1926 to January 1927).
(A) distribution of euphausids in numbers per haul at the surface;
(B) catches of fin whales;
(C) catches of blue whales;
(D) spatial correlation of blue whale catches and the distribution of high inorganic phosphorus (as hatched area) (Hardy and Gunther, 1935).

The same type of work was carried out by the Discovery Investigations on the South Georgia whaling grounds during the twenties and thirties. Figure 11 shows the distribution of *Euphausia superba* Dana in the surface waters north of South Georgia together with the distributions of catches of blue and fin whales (Hardy and Gunther, 1935). The aggregation shown is of the same character and nature as that of herring on to *Calanus* patches. The aggregation–disengagement process must have lasted longer because the position of the total fishery lasted for two months and depended on large and dense fields of euphausids at the surface. A study of causes is shown in Fig. 11d, showing a spatial correlation between the distribution of high phosphorus values in the water and the catches of blue whales.

Blue whales eat euphausids which eat diatoms which consume phosphorus. An inverse correlation might be expected between phosphorus and whales, but it is not so. Here the feeding fishery appears to be connected with details of the production cycle. The correlation observed is really an aggregation of whales on euphausids and an explanation is required of the relationship between euphausids and phosphorus.

Such a correlation was first established by Hentschel and Wattenberg (1930) from the results of the Meteor expedition in the South Atlantic and, more recently, it was confirmed by Reid (1962) for the whole of the Pacific Ocean. Cushing (1971) has explained this correlation by assuming that most of the phosphorus observed in the sea is regenerated by the animals in the zooplankton.

(d) *The albacore fishery in the Eastern Pacific*

In the eastern Pacific the albacore fishery takes place between 170°W and the Asiatic mainland as already described. There are three major zones, the Japanese long line fishery on the equator, that off the Philippines, in the South China Sea and the Dutch East Indies, and that in the Kuroshio extension and eastward to 180°W (Suda, 1963). That in the Dutch East Indies is most prolific in February and there is a fishery in the Kuroshio extension in summer. There is some overlap between the fisheries off the Philippines and those in the Kuroshio extension, forming a fishery south of Kyushu in winter. As shown in an earlier section, fish caught in the Kuroshio extension are largely immature. Spawning takes place in summer in the North Equatorial current towards its eastern end and off the Philippines. The fishery at the same season east of Japan exploits the immature fish which is a feeding fishery. Such albacore fisheries in 1953, 1954 and 1956 spread away from Honshu in the northern drift of the Kuroshio extension as if the fish drifted in the water (Van Campen, 1960). The biomass of immature fish is greater than that of mature fish (Suda, 1963) and the divergences of the Kuroshio extension probably provide more food than those of the North Equatorial current. The complete picture is really the movement of the gyral and the whole migration is structured to provide most food for the immature fast growing fish. The line fishery retreats eastward in early autumn, followed by the fishery south of Japan and north of the Philippines in mid-winter. Then follows that in the Kuroshio extension in early spring and the eastward drift in summer. So between one autumn and the following summer the fishery moves fully half way round the gyral (Nakamura, 1959).

It is not yet known whether the albacore drift with the ocean currents or not. From the recovery of tags, North Sea herring have been shown to drift round the North Sea in the direction of the main North Sea swirl (Høglund, 1955) and from sonar surveys, it has been shown that Norwegian herring move from the polar front north of Iceland to the Faroe Islands in the East Icelandic current (Devold, 1963). If pelagic fish drift with an ocean current, the albacore may move in the direction indicated by the seasonal changes of the fisheries. Then the seasonal eastward drift of the fishery might really indicate the migration of the immature albacore in the richer food areas of the north Pacific anti-cyclone. Between March and May 1954 a number of nuclear bombs were exploded at Bikini Island, which lies in the North Equatorial current in 165°E. Fish caught by the Japanese long liners had to be rejected because of radioactive contamination over rather a broad area for a considerable period of time. During the six-month period after the bomb tests radioactive fish were discarded at sea in the Kuroshio and in the Kuroshio extension as far as 176°W. On straight courses, the distances are enormous, more than 3000 miles in six months; and on a course round the gyre, they would be considerably greater. Fish were also rejected off the eastern coasts of Australia, which were presumably contaminated in the South Equatorial current (Nakamura, 1969).

(e) Conclusion

A feeding fishery depends on fish which gather on to their food patches and which subsequently disengage. It is a dynamic process which is not yet formally described. The North Sea herring fisheries, the Icelandic herring fishery, and the whale fisheries, arctic and antarctic, illustrate the processes and the immature albacore fishery in the Pacific may show the drift of the whole system with the ocean current.

In the great fisheries dependent on food aggregations today, fishermen search for fish with echo sounders and two of the fisheries described in this section were examined scientifically with echo sounders and sonar equipment. Analysis of the processes could be applied to the extensive oceanic fisheries where the food patches must be of enormous horizontal extent; consequently the fishery may endure for periods long enough to be semi-permanent.

In upwelling areas the same result may be achieved for completely opposed reasons. The fish may remain permanently aggregated because food is added to the system as it is grazed. The deep ocean is too big to be sampled quickly with an echo survey, but an upwelling area is of the right size and could be explored in such a way very profitably, as will be shown below.

4. Fisheries in boundary conditions

The boundary is a physical one like a temperature discontinuity or the sharply defined edge of a current system, and fish gather against it. Such boundaries limit the distributions of stocks and usually the cause of aggregation is a combination of physical and biological factors; Harden Jones (1968) has suggested that there is a visible "rheocline" at which fish gather, because they see the moving particles across the boundary.

(a) The Bear Island cod fishery

Cod return from the spawning grounds in the Vestfjord in the West Spitsbergen current and in May and June they are caught on the edge of the Continental shelf west of Bear Island in the Barents Sea. Figure 12 shows the distribution of catches in relation to the bottom temperatures in the first week of June in 1949 (Fig. 12a) and in 1950 (Fig. 12b) (Lee, 1952). In 1949, the cold bottom water (east of the 2° isotherm) extended over the shallow water to the edge of the shelf. The good catches were restricted to a narrow band in another cold water zone lying below 200 fm. In 1950, the 2° isotherm was closer to the island and good catches were found in the same band between the cold water at 200 fm and that on the shelf, but the catches were less dense because the warm water was more widely spread. The concentrations of catches depended closely on the bottom temperature distribution.

The fish may be restricted to the warm current because they have traveled in it or they may return actively from the cold water barrier. If the fish were in the cold water in early summer, their blood content of chloride rose sharply (Woodhead and Woodhead, 1959), but later in the year the fish swam deliberately into it in small numbers. In the early fishery, when the fish have barely started to feed after spawning, their distribution on the Svalbard shelf is sharply limited by the 2° isotherm on the bottom.

Figure 13 shows a more extensive survey made in June 1956 (Richardson et al., 1959). An echo survey and bottom temperature survey was made from south east of Bear Island round the southern edge of the Svalbard shelf and northwards to Spitzbergen. The echo traces were all bottom traces measured within 2 fm of the bottom and were confirmed by trawl catches made at frequent intervals on the survey. Further, a group of trawlers, which were catching cod, were found on each patch of the echo trace. The first survey (Fig. 13a) was made from south to north and it was checked immediately afterwards on the return voyage from Spitzbergen (Fig. 13b). The third survey (Fig. 13c) was made ten days after the first. In both the first and the third surveys on which temperature observations were made, the eastern edge of the echo patch lay between the 1° and 2° isotherm on the bottom.

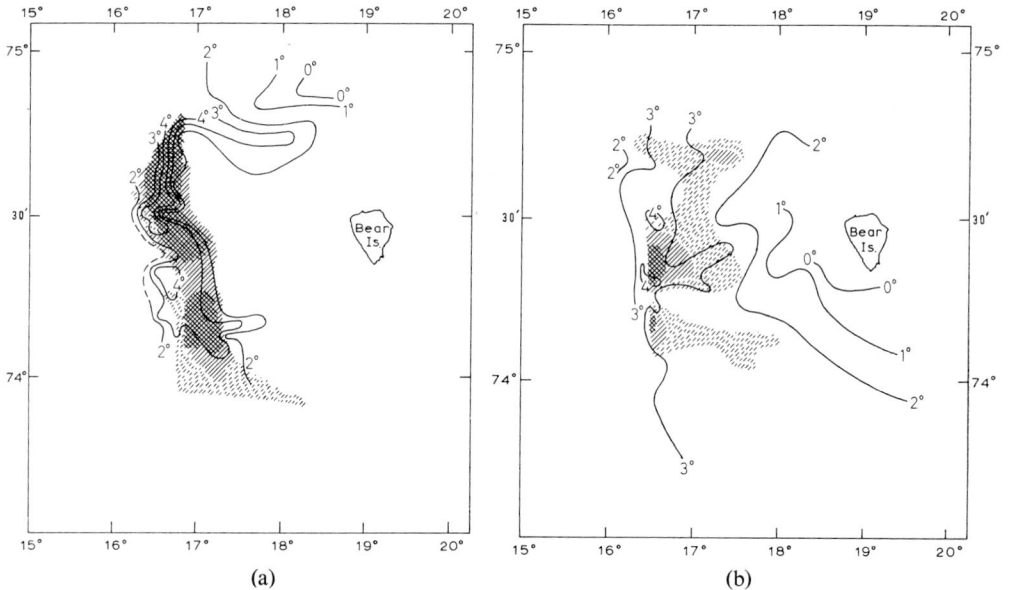

FIG. 12. Catches of cod, as hatched areas, west of Bear Island: (a) in June 1949; (b) in June 1950 (Lee, 1952); the isotherms are those on the bottom.

Comparison of the first and third surveys shows that the fish have moved along the edge of the North West Gully just north of Bear Island and along the southern edge of the Störfjørdrenna, just south of Spitzbergen. The easterly movement was of the order of fifty miles in ten days. It is as if the West Spitzbergen current rolled right up on to the shelf, pressing back the resident Arctic water, and carrying with it the cod population, which had spawned three months before in the Vestfjord.

Echo surveys and bottom temperature surveys were also carried out in July and August 1956 (Beverton and Lee, 1965) and, by this time, the fish had dispersed all over the Svalbard shelf. The 2° isotherm on the bottom limited the distribution of the echo traces between Bear Island and Hope Island (which is some distance to the east). At first the fish were concentrated off Bear Island and then all along the Svalbard shelf and the dynamics of this boundary fishery have been described in pictorial terms. A more rigorous description would first require a fuller physical description of the interaction between the West Spitzbergen current and the cold water on the Svalbard shelf. A fuller description of the cod's behavior would also be needed to determine whether they were thrown on the shelf randomly in the movement of the current or whether they were actively confined within the boundary.

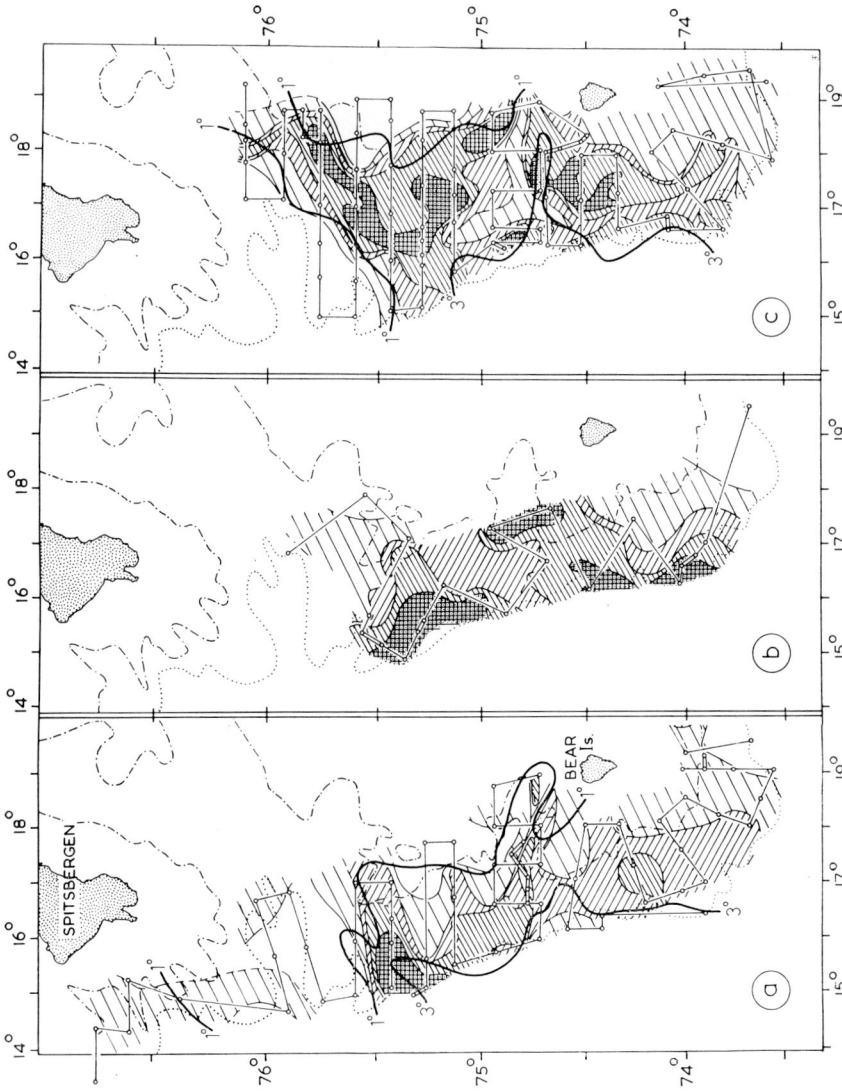

FIG. 13. Echo surveys of cod on the seabed on the Svalbard shelf in relation to the isotherms just off
the bottom:
(a) first survey in early June 1956, north bound;
(b) a check survey, south bound;
(c) third survey, north bound, ten days after the first (Richardson *et al.*, 1959).

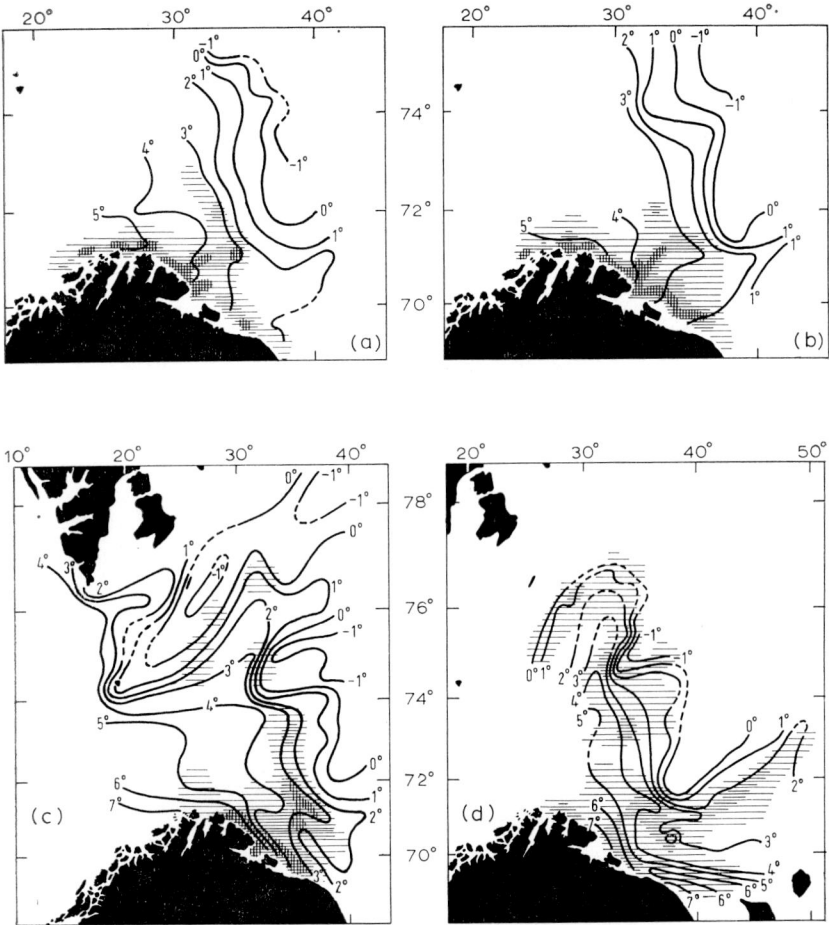

FIG. 14. Echo surveys of cod in mid-water in the eastern Barents Sea, with isotherms at 150 m:
(a) March, April 1959; (b) April, May 1960; (c) September, October 1958; (d) September, October
1959 (Hylen *et al.*, 1961).

(b) *The North Cape fishery*

Echo and temperature surveys have been combined in another study of a fishery on the
Arcto-Norwegian cod stock. The isotherms in Fig. 14 are at a depth of 150 m and the fish
densities are estimated from counts of individual fish in mid-water, almost certainly cod
(Hylen *et al.*, 1961). Surveys in March and May are shown at the top of Fig. 14a and b;
surveys in September–October are shown in the bottom of Fig. 14c and d. Fish are restricted to areas warmer than 2°C except in two areas in the Central Barents Sea. Between
March and May, fish have drifted eastwards from the Lofoten spawning grounds in the
North Cape current. In September and October, the distribution of echo traces is again
limited by the 2° isotherm, but is strikingly different in the two years represented, 1958
and 1959. Comparison of the four charts suggests that the 2° isotherm "opens" the Eastern
Barents sea to the cod's migration in spring and "closes" it again in autumn. Expressed in
another way, Midttun (1964) found a positive correlation of Norwegian spring catches of

cod with temperature and Konstantinov (1964a, b) showed that Murman and Finmark catches are inversely correlated.

The British and Norwegian studies in the Barents Sea are complementary in showing that the 2° isotherm is effective in spring, summer, and autumn, in mid-winter and on the bottom. The isotherm is a true boundary to the distribution of the Arcto-Norwegian cod and it is likely that the fish respond actively to the temperature barrier. On the Svalbard shelf in June the temperature discontinuity can be as sharp at 1°C in about 2 km (Lee, 1952). Fish which can detect a temperature difference of 0.03°C (Bull, 1936) might well respond actively to such a boundary, because the threshold of detection can be reached in a straight course of 100 m or in 50 sec for a cod of average size (70 cm) swimming at three lengths sec^{-1}.

(c) *The summer fishery on herring by the Russians in the Norwegian Sea*

The Norwegian herring move away from their spawning grounds into the Norwegian Sea as described above. Between Iceland and Jan Mayen in summer, herring gather and feed against the polar front, and, since 1953, up to six hundred Russian drifters have worked in summer. They also work in the area north of the Faroe Islands in spring and early summer. It is likely that the polar front fishery depends upon the generation of heavy production in the stable layer of melt water near the barrier (Marshall, 1958) but such an explanation is not valid for the Faroe Island fishery.

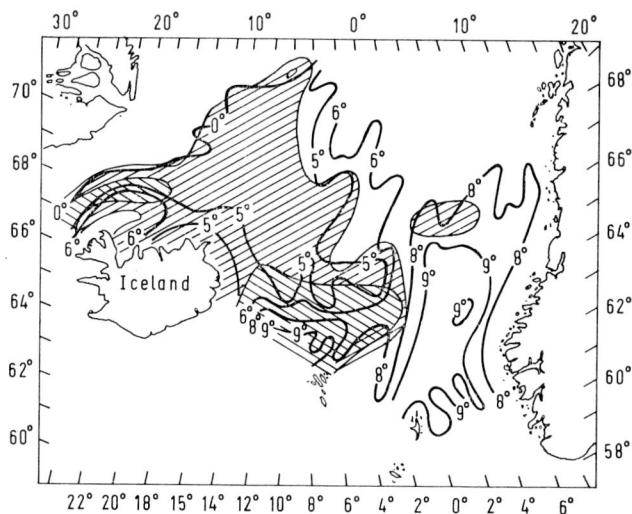

FIG. 15. A sonar and echo survey of the Norwegian Sea in June 1955. Echo traces and sonar traces are shown near the polar front between Jan Meyen and Iceland. They are also found where the isotherms crowd, north east of the Faroe Island, where the East Icelandic current meets the warmer Atlantic stream (Tåning *et al.*, 1957).

Figure 3 shows the structure of the summer fishery from Devold (1963) with the East Icelandic current, shown diagrammatically as a tongue pointing south-eastwards from the polar front between Iceland and Jan Mayen, indicated by the 0°C isotherm at the surface. Figure 15 shows a distribution of sonar and echo traces and the distribution of the surface temperature in the summer of 1955. Echo traces, indicated by the 5° isotherm at the surface, were found close to the cold water, north and north east of the Faroe Islands, where a fishery also takes place. In both regions the fishery occurs where the isotherms crowd. However, the important point is that the crowding in miles is not very marked and the temperature differences per mile cannot be perceived by the fish at all. Yet there might be sharp undescribed boundaries within the isotherms.

Temperature boundaries like those shown in Fig. 15 can arise from the conflict of water masses. The concentrations of fish and, hence, fisheries, occur where the boundaries are sharpest and where the currents must be relatively faster than elsewhere in the ocean. The boundary north west of the Faroe Islands is quite different from the possibly lethal one between Iceland and Jan Mayen. It is a dynamic boundary because the crowding of isotherms indicates water masses which flow relative to each other. Fisheries often occur where such water masses conflict. The cause of concentration is sometimes attributed to a mixing of the waters generating nutrients at the surface and hence food, but in the Norwegian Sea, more food is found in the center where isotherms are far apart (Pavshtiks, 1960). The reason why fisheries occur where the isotherms crowd remains unknown.

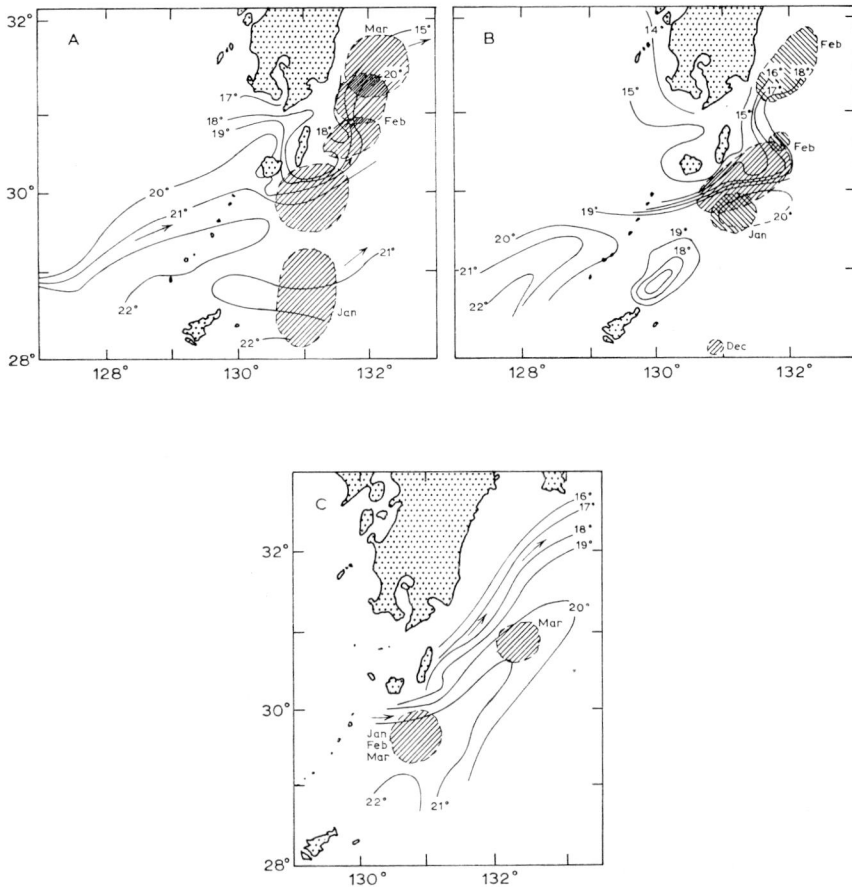

FIG. 16. Distribution of bluefin catches in the Kuroshio current; the water temperatures were recorded at 100 m: (a) spring 1937; (b) spring 1940; (c) spring 1941 (Uda, 1952).

(d) *The bluefin tuna in the Kuroshio current*

Between mid-January and March the bluefin tuna fishery (*Thunnus thynnus* Linnaeus) moves from south of the island of Kyushu in southern Japan along the Kuroshio current to an area west of that island. Figure 16 shows the patches of the bluefin tuna between January and March in 1937, 1940 and 1941 (Uda, 1952) as shown by catches. The temperature distribution is drawn from samples taken at 100 m and the currents (shown by arrows) were estimated at the surface. The fisheries from January to March are found on the western boundary of the main Kuroshio current. The gyral concentrates in velocity at its western boundary, tuna are concentrated there, and a fishery by the Japanese fishermen occurs there; it is a region of high divergence for the tuna may aggregate in productive waters. It is also possible that the fish concentrate at Harden Jones' "rheocline" (1968).

Uda (1952) also shows that the position of the bluefin fishery varies from year to year. Figure 16c shows the distribution of catches in 1941 when the axis of the Kuroshio was in its "normal" position. Figures 16a and b show the catches in 1937 and 1940 when the axis of the Kuroshio was diverted southwards by a cold water mass from the north. Then the fisheries were concentrated and high catches were made, so the availability of fish to capture can be modified directly by shifts in the boundary of a major current.

An interesting phenomenon associated with the current boundaries is the "Siome". Uda (1937) wrote: "on August 14, 1937, passing through south off Kusiro, we observed a 'Siome' accompanied with ripping white waves and shoals of skipper with associates of sea birds near by". A sharp fall of water temperature of about 5°C was observed on the surface thermograph. This was the very edge of the current system and the Japanese fishermen used it to find fish. The boundary rumbles and hisses and these noises can be heard from the deck of a searching ship. The "siome" is quite sharp enough to be easily detected by fish. At such a boundary, the current is swiftest but there is no reason why this in itself should generate aggregation unless there is a rheocline.

(e) *The Baltic outflow fishery for herring*

The Baltic outflow into the North Sea runs from the Skagerrak along the Norwegian coast and above the Norwegian deep water. Figure 17 shows a typical section across the western edge of the Norwegian deep water, showing the less salty and rather cooler Baltic water lying on the surface over the deep water. Figure 18 shows the positions of echo traces of herring along the temperature boundary between the cool Baltic outflow and the warmer North Sea water. The echo traces were identified by herring catches made by drifters in the same area. This account is taken from Steele (1961).

The herring are feeding at this time. The herbivorous copepods live in equal numbers on either side of the boundary but there is more chlorophyll in the fresh Baltic outflow and more euphausids. The euphausids probably live in the warmer water below the Baltic outflow and climb into the cold water at night to feed and the herring feed on the euphausids. The boundary changes its position very quickly and the herring, as represented by the echo traces, tend to follow it. They do not move eastward into the Baltic outflow but remain in the warmer water on the edge of the Norwegian deep water. The euphausids are found at the surface at night in the cooler water of the Baltic outflow. Herring below the western edge of the outflow rise to the surface at night to feed.

In the winter and early spring, the waters of the North Sea are turbulent and the chance of a plant cell remaining in the sunlit layer is low. It is the ratio of the depth of the sunlit layer to that of the mixed layer which governs the onset of production in the sea. So long

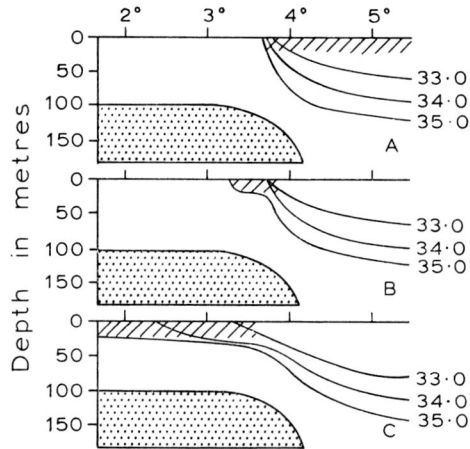

FIG. 17. Sections across the Baltic outflow in March, shown as isohalines. The three sections show the westward shift of the boundary; the productive layer is hatched (Steele, 1961).

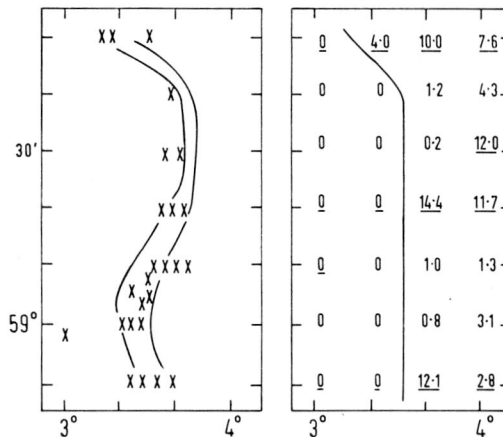

FIG. 18. The western boundary of the Baltic outflow; the full lines mark it and the crosses show the positions of the echo traces. On the right are shown the catches of euphausids on both sides of the boundary (Steele, 1961).

as the sea remains turbulent, production is inhibited. In the Baltic outflow, the surface water cannot be overturned, because it is light being less salty, yet cooler. As the depth of the layer is about that of the sunlit layer, production starts very early and much earlier than in any other part of the northern and central North Sea.

Overwintering herring live on the western edge of the Norwegian deep water (Anon., 1965), between the Skagerrak and the Viking Bank. The boundary provides a mechanism by which the fish can get food earlier than in any part of the North Sea north of the Dogger and it is really less important than the production which takes place to the eastward in the Baltic outflow. Yet the fish remain close to it, apparently to their advantage.

(f) *Conclusion*

For the biologist an attractive range of problems has been studied, because the biological and physical factors are so profoundly intermixed. In the Bear Island cod fishery, the fish aggregate at a front that moves up and across the Svalbard shelf. A similar boundary exists off the North Cape, a gate through which the cod may pass in spring on their way to the south east Barents Sea. The fish may actively turn from these boundaries, because they can detect the cold water if the boundary is sharp enough. In the late autumn the fish return westward through the gate on their way back to the Vestfjord. Physiological factors appear to be absent in the summer fishery for herring in the Norwegian sea. The arrival of bluefin tuna off the island of Honshu is associated with the position of the Kuroshio current. Fish may drift with the currents and where they are fastest, at boundaries, they will move faster. But sometimes the fish appear to aggregate at a boundary itself.

The Baltic outflow fishery for herring is a complex one. The physical structure is fairly simple, a stable layer lying near the surface alongside mixed water. The productive cycles differ on either side of the boundary, euphausids climbing into the Baltic water masses. As a consequence, the herring stocks gain great advantage from the food at this boundary, on the edge of their overwintering area.

5. *Fisheries in the upwelling areas*

In an upwelling area, cold water from the depths reaches the surface under the stress of winds parallel to the coast, particularly in the eastern boundary current. The California current runs southward at the surface in the main direction of the wind. By Coriolis force, the southerly wind generates an offshore movement of water, which in its turn drags water up from about 200 m; the upwelling process carries the water to a dynamic boundary about 100 km offshore. In seasons of upwelling a counter-current below 200 m replaces the water which has drifted offshore. The cold water brought to the surface generates fog, and the land is dry because of the longshore winds. Because the water wells up from the depths it is rich in all the nutrients and throughout the world the presence of high nutrients is associated with the presence of upwelled cold water and high quantities of plankton and of fish. Much of the information used in this section is summarized in Cushing (1971).

(a) *The sperm whale fisheries*

The whalers from Nantucket covered the whole world on their long voyages. They recorded the positions of their catches in their log books and Fig. 19 shows the positions of capture of 36,908 sperm whales caught between 1729 and 1919 (Townsend, 1935). Figure 19a shows the positions from April to September (summer in the north and winter in the south) and Fig. 19b shows them from September to April (winter in the north and summer in the south). The real and astonishing point about this figure is that it shows, amongst other things, the main centers of upwelling in the world, the Peru current off the western coast of South America, the Benguela current off the western coast of South Africa. It shows areas between Morocco and Dakar, off southern Arabia, off California, and elsewhere. The most famous ground of all, the Line in the Pacific, is an upwelling area of a special type.

Sperm whales were also caught in summer in the Sargasso Sea and in its oceanic analog, west of Japan, in the central north Pacific. They were also caught in summer off Brazil. Upwelling does not take place in any of these three areas, where nutrients are so low that

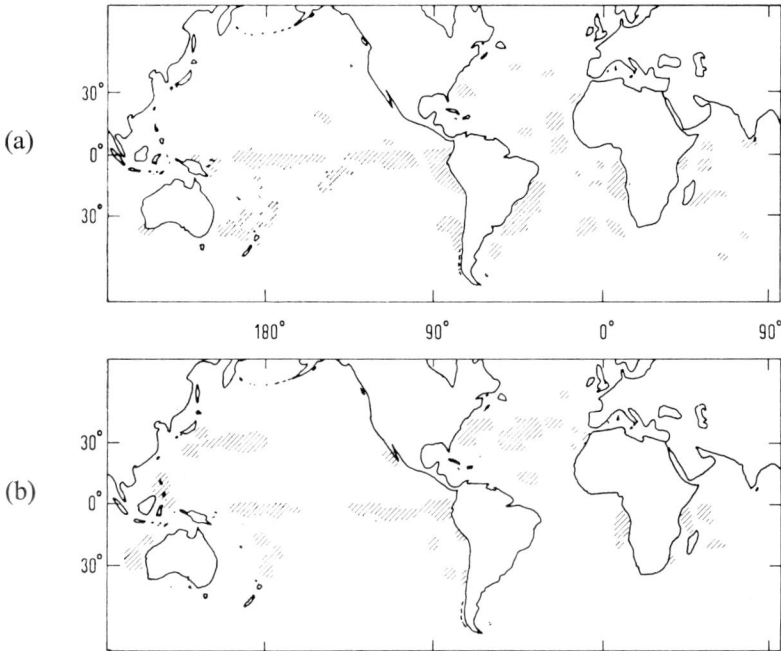

FIG. 19. Positions of capture of sperm whales throughout the world by Nantucket whalers: (a) April to September; (b) October to March (Townsend, 1935).

they are near the limits of detection. The plankton is very thin and fish are only caught by scientists. But sperm whales eat fish and so it is possible that the simple association between fish and nutrients in upwelling areas is spurious, as suggested in an earlier section.

(b) *The tuna fishery along the Pacific equator*

The yellowfin fishery (*Thunnus albacares* Bonnaterre) on the Pacific equator was exploited first by the Japanese working eastwards along the Line from the East Indies and then by Americans working from Hawaii in exploratory fishing. Yellowfin are also exploited by Californian fishermen working off California, Mexico, and Central America. It is an enormous area, 20° in latitude to 120°W and the catch ranges from $3\frac{1}{2}$–$11\frac{1}{2}$ tuna per 100 hooks, and the best catches are made in September (Calkins and Chatwin, 1967). From October to June, half the fish appear to be spent but in September the proportion is about one fifth and the fish appear to be feeding. Perhaps spawning and feeding goes on all the time, with perhaps a little decrease of spawning intensity taking place in September.

The distribution of the catch per 100 hooks across the Line at intervals of about 1° at 160–170°W shows that the best catches are found from the equator to 2°N. There are also quite good catches found at 5°S (Murphy and Shomura, 1955). The band of high catches just north of the Line is also a band of high zooplankton in a region of divergence (Fig. 20). From south to north (or left to right) the figure is divided into labeled sections, representing the different currents in the complex equatorial system (King and Hida, 1954); SEC is the westbound South Equatorial current; NEC is the westbound North Equatorial current; CC is the eastbound Counter-current; DIV is the zone of divergence; and CONV is the

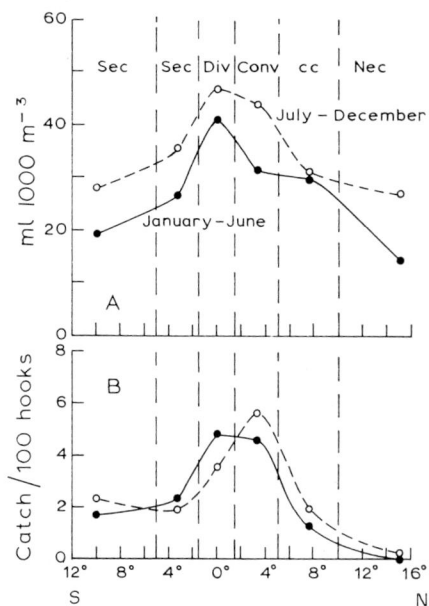

FIG. 20. Distribution of zooplankton (in ml 1000 m⁻³) and tuna in numbers per 100 hooks in the Pacific equatorial current complex in two periods, January to June and July to December; see text for explanation (King and Hida, 1954).

zone of convergence. From July to December, a period of south east trades, the largest quantities of yellowfin were found in the zone of convergence; from January to June, a period of north easterly or variable winds, the fish were found both in the zone of divergence and that of convergence. The peak of zooplankton volumes is found all the year round in the zone of divergence. Seasonal differences are not great. From 180 to 120°W along the Line, zooplankton abundance is more than doubled, a difference of roughly the same order as that across the current complex.

The Equatorial current system is the boundary between the north and the south sub-tropical gyrals. The zone of convergence is one where the water is forced to sink on the southern boundary of the counter-current. The zone of divergence is that where water is brought to the surface from about 100 m on the northern boundary of the South Equatorial current. Thus the divergence is a form of upwelling, bringing cold water towards the surface. The higher quantities of nutrients are said to generate larger quantities of zooplankton in the zone of divergence than elsewhere. Tuna aggregate here, feeding on the fish that feed on zooplankton. The zone of divergence can be identified as cooler water at the surface. Yellowfin catches north and south of the Line at 140°W have been compared with the temperature distribution in depth. Sometimes they are concentrated on the northern half of the zone of divergence. If the cooler water is spread over a wide area, the tuna do not concentrate.

It is unlikely that the fishery for yellowfin along the Line depends only on the zone of divergence, which is the main zone of upwelling in the Equatorial complex. The difference in zooplankton density between the zone of divergence and outside it is not very great, but there can be a ten-fold difference in yellowfin catch. A simple aggregation of herring on to a *Calanus* patch in the North Sea requires much bigger food differences to generate such a catch difference so it is likely that other mechanisms are at work.

(c) *The pilchard and maasbanker fishery in the Benguela current*

The fishery is concentrated mainly in two regions, St. Helena Bay in the Union of South Africa and Walvis Bay in S.W. Africa. Up to 400,000 tons of pilchard (*Sardinops ocellata* Pappé) and 70,000 ton of maasbanker (*Trachurus trachurus* Linnaeus) were landed from St. Helena Bay and up to 600,000 tons of pilchard from Walvis Bay in the mid-sixties; more recently, catches in St. Helena Bay have declined and those off S.W. Africa have increased. In St. Helena Bay the fish are caught in the late autumn, March to June.

The mechanism of upwelling is shown in Fig. 21. The general northward movement of deep water and coastal water is called the Benguela current. There is a compensatory current in the deep water close to the shelf and the upwelling is generated by S.S.E. or E.S.E. winds (Hart and Currie, 1960). These offshore winds drive the surface waters away from the shelf, drawing cold water from the depths toward the coast. The upwelling is probably centered in three regions (a) from St. Helena Bay to the mouth of the Orange River on the boundary between S. Africa and S.W. Africa; (b) from Lüderitz to Walvis Bay, a distance of about 150 miles; and (c) from Cape Frio to the Cunene river mouth, on the Angolan border, a distance of about 100 miles.

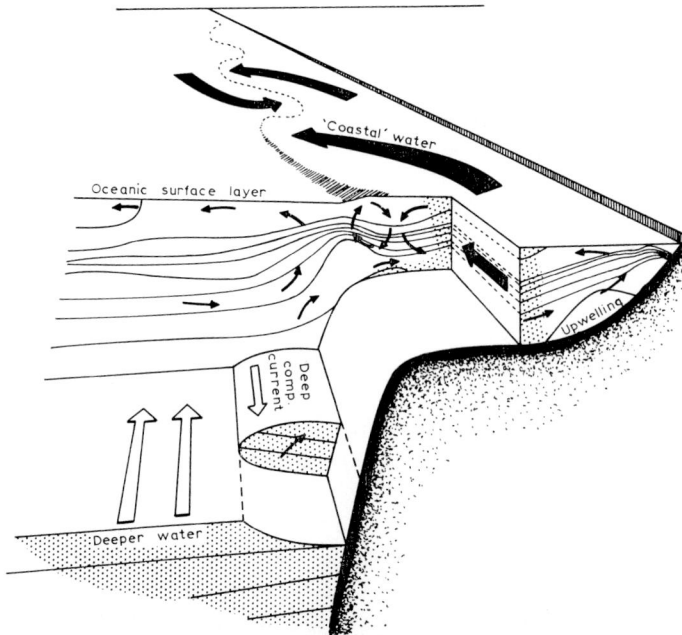

FIG. 21. The structure of the Benguela current and the upwelling system (Hart and Currie, 1960).

The season of upwelling extends from October to February in the south and the greatest phytoplankton production occurs between September and December (Hart and Currie, 1960); and there is ample food available for larval pilchards. But between Walvis Bay and the Orange River the season may be much more extensive, perhaps lasting most of the year, with an intense period in mid-winter. As in the other main upwelling regions in the world there is a poleward movement in intensity as the season progresses, presumably associated with an intensification of the South Atlantic high. There is a guano industry in the Benguela partly on artificial islands as in Walvis Bay, and partly on real islands. Penguins, gannets, and cormorants eat 48,000 tons of pilchard and 7000 tons of maasbanker in St. Helena Bay (Davies, 1957). The Benguela current yields about 1 M tons of pelagic fish and the stocks were possibly under-exploited until very recently. There are stocks of hake (*Merluccius productus* Ayres) available to the edge of the shelf which have been investigated by echo survey as will be described later.

(d) *The Californian sardine fishery*

The sardine fishery (*Sardinops caerulea* Girard) failed in the late forties and has not since recovered. At its height the fishery was based on three areas, the north west Pacific, San Francisco-Monterey and San Diego. Biological differences have been found between the three groups (Sprague and Vrooman, 1963) and fish tagged in the south were recovered in the south. The failure in the fishery was progressive from north to south. During the fifties the spawning areas were found between Point Conception and the coast of Bahia California. Charts of egg surveys (Fig. 22) may be regarded as stock quantities because the survey cruises were carried out continuously throughout the spawning season, January to July, with a peak in April and May. It is likely that the spawning grounds were roughly in the same position from 1930–50 but more extensive in area and that the fish

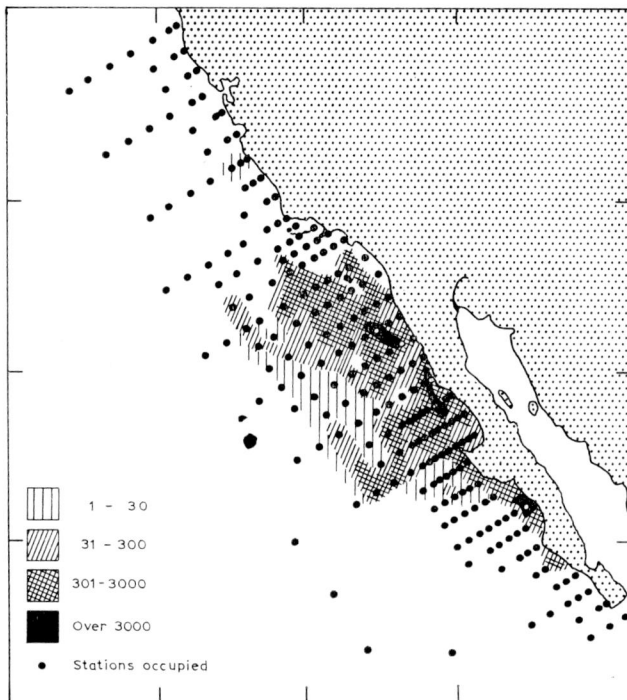

FIG. 22. An egg survey (numbers/(10 m)^3) off the coasts of California and Mexico (Ahlstrom 1966).

migrated northwards after spawning, forming a fishery in summer and in late summer from
San Diego to British Columbia.

As in the other main upwelling areas, longshore winds blow towards the equator and
bring cold water up from the depths to the surface inshore. The upwelling patches tend to
break inshore of the south bound Californian current. Figure 23 shows the centers of upwelling off Oregon, south of Monterey, south of Point Conception and off Punta San
Eugenio. The upwelling area has extended southward since the collapse of the fishery in
1949 as shown in the figure. Although upwelling is going on all the time, it reaches a peak
in April and May which is the peak spawning season. Spawning during the fifties was
associated with the upwelling around Punta San Eugenio and Punta Bahia and so it appears
that the sardines spawn at the points of upwelling.

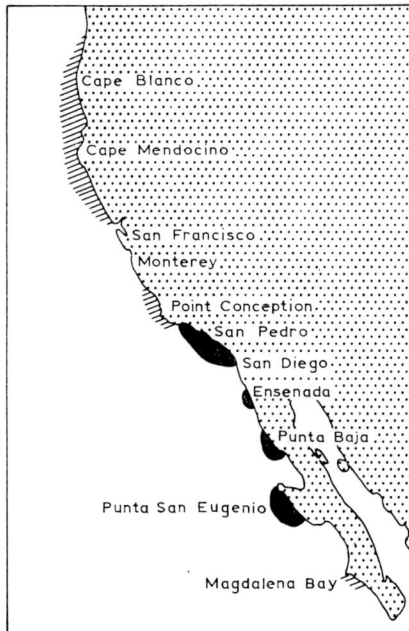

FIG. 23. The regions of the Californian sardine fishery before its collapse in the late forties. The hatched
patches show the region of upwelling before 1949; the black patches show regions of upwelling before
and since that date (Calcofi Progr. Rep. 1952–3).

5. The Dome fishery for tuna off equatorial America

Yellowfin and skipjack (*Euthynnus pelamis* Linnaeus) tuna are caught between 35°N
and 20°S on the Pacific coast of the Americas. A very productive area is that off the coast
of Central America which is described in oceanographic terms as the Costa Rica Dome,
where the Equatorial counter-current turns north and then west into the North Equatorial
current. In this area the thermocline is quite close to the surface (Wooster and Cromwell,
1958) and the region is biologically productive (Holmes *et al.*, 1957; Brandhorst, 1958).
Cromwell (1958) suggested that the productive mechanism is mainly generated by the
swirling action of the counter-current. Obviously such a mechanism would be most effective
when the thermocline is nearest the surface as it is in the Costa Rica Dome.

6. The anchoveta fishery off Peru

The largest fishery in the world takes place off the coasts of Peru in the region of most intense upwelling in the Peru current. About 9 or 10 M ton of anchoveta (*Engraulis ringens* Jenyns) are caught there each year. The most intense period of upwelling occurs in mid-winter, but there is probably some upwelling all the year round. An interesting point is that the zone of biological production extends far to seaward (Schaefer *et al.*, 1958), beyond the dynamic boundary at 100 km and beyond the fishery. From surveys of egg production, the anchoveta outnumbers all other species at all seasons by a factor of ten (Flores *et al.*, 1967). The egg distributions, like those off California, show that the fish spawn close to the shore at the point of upwelling. The fish, like other anchovies, have phytoplankton in their guts, but they feed on zooplankton and in fact in area the distributions of anchoveta eggs exclude those of zooplankton. The inverse correlation in space is not exact, which suggests that there may be a time lag in its generation. But it would appear that the anchoveta are numerous enough to strip the zooplankton animals from the water (see Cushing, 1971, for a brief account of the fishery).

Figure 24 shows the distribution of eggs associated with that of the zooplankton (Flores *et al.*, 1967). The vertical distribution of echo traces shows that the fish live in the upper part of the thermocline. The biological structure of this upwelling area is comparatively simple; anchoveta eat zooplankton and some phytoplankton and are themselves eaten by the guano birds and are fished by man. Up to 2–3 M ton per year can be taken by the guano birds. The food is simple and the use of echo survey horizontally and vertically allows a sampling system to be exploited most efficiently. However, there are complicating factors: periodically in the summer time, at Christmas, a surface of warm water floods in from the north. It is called El Niño (the little boy, because it happens at Christmas time) and the fish go below it and the guano birds are deprived of food; they starve or fly south. The fishery stops and rotting plankton and dead fish are deposited on the beaches. The warm surface layer stops the upwelling process and the whole biological unit goes short of food (Posner, 1957; Bjerknes, 1961).

Conclusion

The distribution of the sperm whale fisheries was used to chart the areas of upwelling. In such areas, cool water is brought to the surface and where this happens, plankton quantities are high. Fisheries develop in these regions and fish spawn where there is much food available for the larvae. The fluid mechanisms are three-fold, pure upwelling as in the Benguela current or off the Californian coast, divergence as in the border of the Equatorial counter-current, and advection as across the top of the Costa Rica Dome.

Fisheries in the main areas have been briefly examined, that for tuna with pelagic long lines in the equatorial complexes, that for pilchards with purse seines in the Benguela current off South Africa, that for sardines with purse-seines in the California current, that for tuna in the Costa Rica Dome and that for anchoveta with purse-seiners in the Peru current. All of them depend upon upwelling processes in one way or another. An upwelling area usually has a complex of currents within it by which fish such as hake or sardines can retain their geographical position without being drifted away into mid-ocean, so long as the fish leave or enter a current by sinking or rising. The food chain structure, like those in high latitudes, tends to be simple and therefore can be examined by fairly simple sampling techniques.

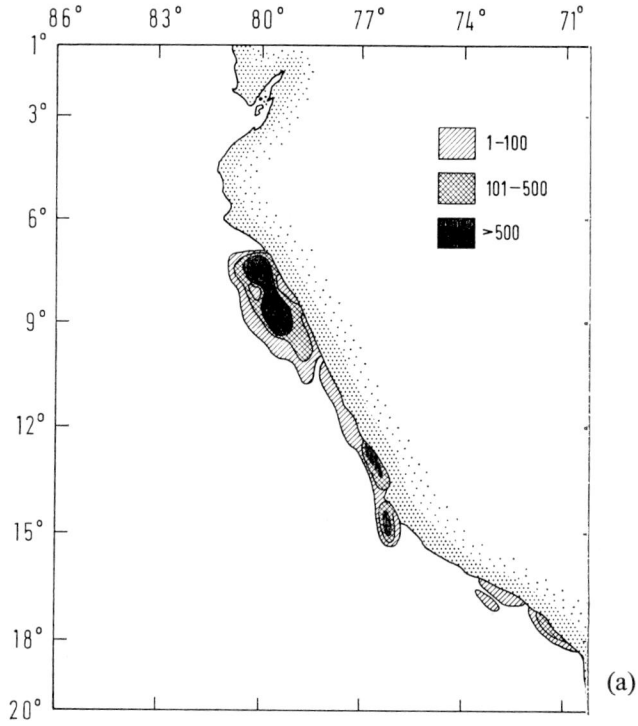

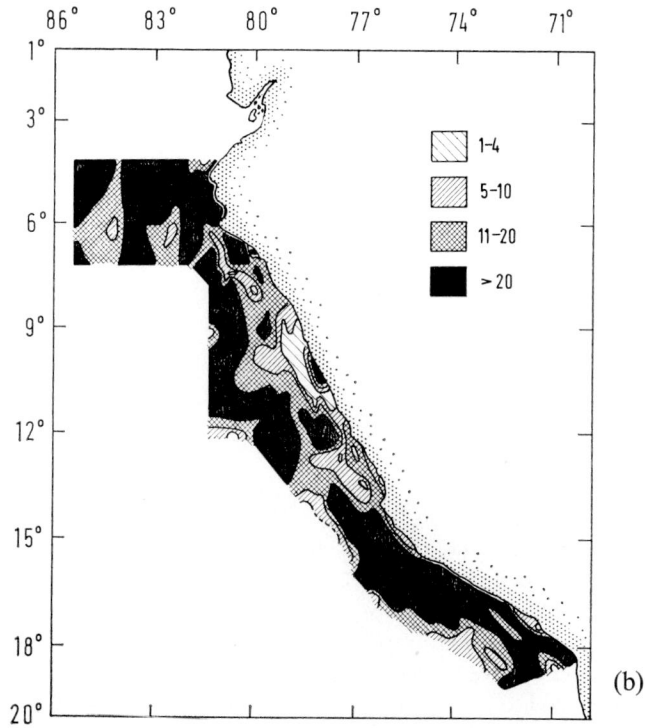

FIG. 24. The distribution of (a) anchoveta eggs off Peru and (b) that of zooplankton (Flores *et al.*, 1967).

Echo survey techniques have been used in Peru with some success for charting the distribution of pelagic fish. Demersal fish, like hake, have also been studied by these methods off Peru and South Africa, as will be shown later. But the study of an upwelling area is really that of a slow and complex horizontal movement, with different components vertically. The traditional methods using fisheries statistics can reveal differences in concentration of catch in small units of time and space. But we need to add to this study the details of distribution horizontally and vertically as revealed by acoustic technique in relation to environmental factors.

7. The evaluation of resources by exploratory fishing

Until World War II most exploratory fishing was carried out by fishermen; the exploration of the North Sea by smacksmen and trawlermen (Cushing, 1966) and the exploration of the halibut grounds in the north east Pacific are traditional examples of such exploration. Fisheries biologists have conducted some surveys, but usually exploratory surveys have been executed by sponsored fishing vessels. Two exceptions are provided by the exploration by trawl of the north east Pacific and the Guinean trawling survey. It might be considered that such careful and meticulous work, as will be described, has been outdated by newer methods of acoustic survey, but this is not so. Acoustic methods cannot be used to identify fish and where there are many species the acoustic survey can best be used to support the trawl survey and to execute it more quickly. This section describes the form of information which is often needed by fisheries biologists and underlines the fact that acoustic survey is no substitute for it. For this purpose, all such a survey can do is accelerate the procedures by making it possible to sample the areas where fish are abundant more quickly.

(a) The north east Pacific

The study goes back in history to the early explorations made by the fishermen, but it is essentially one made by the vessels of the U.S. Bureau of Commercial Fisheries and has been reported in some detail by Alverson et al. (1964). The Albatross worked between California and the Bering Sea between 1889 and 1921: in 1940 and 1941, the Alaska king crab was investigated with standard commercial gear. From 1948 onwards, exploratory surveys were carried out with beam trawls, otter trawls, and Gulf shrimp trawls, by the Bureau of Commercial Fisheries and the International Pacific Halibut Commission. During the period 1755 hauls were made with otter trawls, 530 with Gulf shrimp trawls and 339 with beam trawls. They were distributed in depth of the states of Washington and Oregon and Alaska (on the southern coasts, in Bristol Bay, and in the Chukchi Sea).

Table 1 summarizes the results obtained in lb hr^{-1} trawled by 50 fm depth sectors for the more abundant species of flounders, rock fish, round fish, elasmobranchs, and ratfish. The areas sampled are classed into three areas in the Gulf of Alaska (Shelikov Strait is inside Kodiak Island and Prince William Sound is not far from Cook's inlet), the coast of Oregon and Washington, and the Strait of S. Juan de Fuca between Vancouver Island and the United States.

The flatfish were classed as coastal species (0–20 fm), shallow species (100–149 fm), and deep water species (150–500 fm).

In Table 2, only the Alaska plaice (Pleuronectes quadrituberculus Pallas) is classed as a shallow species; all the rest, including the Pacific halibut, which is not well sampled by the trawl, are classed as deep species. It will be seen that the flounders are not common

TABLE 2. CATCHES (lb hr⁻¹) TRAWLED IN DEPTH ZONES IN DIFFERENT REGIONS[a] (Alverson et al., 1964)

Species	Oregon-Washington										Alaska														
	Outside					Strait of S. J. de Fuca					Gulf of Alaska					Prince William Sound					Shelikov Strait				
	1-49	50-99	100-149	150-199	200-299	1-49	50-99	100-149	150-199	200-299	1-49	50-99	100-149	150-199	200-299	1-49	50-99	100-149	150-199	200-299	1-49	50-99	100-149	150-199	200-299
FLOUNDERS																									
Atherestes stomias	0	159	328	98	50	10	82	80	—	—	78	292	323	98	50	3	87	40	0	51	—	118	180	110	—
Microstomus pacificus	0	145	494	402	252	t	1	t	—	—	t	4	23	105	473	0	0	t	0	3	—	2	t	0	—
Glyptocephalus zachirus	0	49	41	26	24	t	1	1	—	—	3	11	24	46	63	t	t	t	0	t	—	3	t	0	—
Hippoglossoides elassodon	0	t	t	t	0	26	t	t	—	—	16	83	67	62	0	24	58	6	0	11	—	254	243	35	—
Limanda aspera											2	2	0	0	0	6	t	t	0	0	—	2	0	0	—
Pleuronectes quadrituberculus											t	t	0	0	0	11	2	0	0	0	—	6	t	0	—
ROCKFISH																									
Sebastodes alutus	0	34	1693	609	170	0	3	21	—	—	1	154	374	194	t	0	11	2	0	t	N	6	17	4	N
Sebastodes flavidus	0	29	2	0	t	1	t	0	—	—															
Sebastodes pinniger	0	15	19	t	0	1	t	0	—	—															
Sebastolobus spp.	0	1	15	23	68	0	0	0	—	—	t	1	39	19	102	0	0	0	0	2	N	0	t	0	N
ROUNDFISH																									
Gadus macrocephalus	0	2	1	0	0	170	71	43	—	—	97	67	25	8	t	8	60	14	50	t	—	36	25	18	—
Theragra chalcogramma	0	1	4	5	0	1	17	19	—	—	5	82	73	28	0	2	114	34	0	72	—	39	48	1	72
Anoplopoma fimbria	0	46	273	261	476	1	2	1	—	—	5	25	60	14	284	2	t	t	t	7	—	13	8	10	—
Merluccius productus	0	269	25	118	35	0	t	t	—	—	3	6	5	t	0	0	0	0	0	0	—	0	0	0	—
Ophiodon elongatus	5	11	7	1	0	2	15	66	—	—	t	t	t	0	0	0	0	0	0	0	—	0	0	0	—
ELASMOBRANCHS AND RATFISH																									
Squalus acanthias	0	33	100	24	2	34	388	330	—	—	47	7	3	3	0	8	21	6	0	51	—	6	1	0	—
Raja spp.	24	14	18	12	21	49	43	69	—	—	94	28	37	17	32	t	8	15	0	76	—	24	4	25	—
Hydrolagus colliei	0	7	10	17	2	71	477	612	—	—															

[a] t = trace;   N = number of hauls made.

in the Strait of S. Juan de Fuca or in Prince William Sound. The turbot, or arrow tooth flounder (*Atherestes stomias* Jordan and Gilbert) and Dover sole (*Microstomus pacificus* Lockington) are common off Washington and Oregon and in the Gulf of Alaska; the turbot is quite common in the Shelikov Strait. The flat head sole (*Hippoglossoides elassodon* Jordan and Gilbert) is the commonest flatfish in Shelikov Strait and is more abundant there than elsewhere. The Pacific halibut (*Hippoglossus stenolepis* Schmidt), not shown in Table 2, was most abundant in the Gulf of Alaska and along the Alaskan peninsula, at a rate of about 80 lb hr^{-1} trawled in depths less than 50 fm and at about 50 lb^{-1}hr in depths between 50–100 fm. The rockfish live in deeper water, so are virtually absent from Shelikov Strait, Prince William Sound, and the Strait of S. Juan de Fuca. Of the species in Table 2, the Pacific Ocean perch (*Sebastodes alutus* Gilbert) is much the commonest for it is found in considerable quantities in 100–149 fm off the coasts of Oregon and Washington; in the Gulf of Alaska at the same depth range, the density is only about one quarter of that off Oregon and Washington. The roundfish, or gadoids, are spread in depth. The Pacific cod (*Gadus macrocephalus* Tolesius) is found in shallow water in the Strait of S. Juan de Fuca and in the Gulf of Alaska. The pollack (*Theragra chalcogramma* Pallas) is limited to moderate depths in the Gulf of Alaska and is not very abundant. The black cod (*Anoplopoma fimbria* Pallas) is abundant in deep water and moderate depths off Oregon and Washington and is limited to deep water in the Gulf of Alaska. The Pacific hake is only found in shallow water (50–99 fm) off Oregon and Washington, as might be expected for a generally southern species; in any case it spawns off southern California. The elasmobranchs and ratfish were most abundant in the Strait of S. Juan de Fuca, between 50–144 fm, but they are not very abundant elsewhere.

In the table are examined the demersal fish from two oceanic structures, those from the California current off the coasts of Oregon and Washington and that from the Alaska gyral in the Gulf of Alaska. The Pacific hake is restricted to the southern area and the Dover sole and black cod are found in deeper water in the Alaska gyral as opposed to the Californian current. These results were obtained over more than a decade and represent an enormous amount of work. Much of it could not be executed by acoustic survey because flatfish of moderate size cannot be detected as distinct from the seabed. But acoustic survey could be applied to the rockfish and roundfish stocks; indeed Alverson (1967) has applied such methods to the stock of Pacific hake off the coasts of Oregon and Washington, as will be described later.

(b) *The Guinean trawling survey*

In 1960, the Commission for Technical Cooperation in Africa (CTCA) decided to establish a trawling survey on the Continental Shelf between Mauritania and Angola. The operational phase was executed in two parts, from August to December 1963 and from February to June 1964; three trawlers were used, with scientists from France, Ivory Coast, West Germany, Ghana, Netherlands, Belgium, U.K., U.S.A., and F.A.O. Fisheries Division. Between Cape Roxo in Gambia and the Congo, sixty-three sections were worked which were grouped in thirteen statistical areas. Hauls were made at eight stations on each transect at the following depths, 15/20 m, 30 m, 40 m, 50 m, 70/75 m, 100 m, 200 m, and 400/600 m. A grab or dredge was hauled at each station, together with certain oceanographic observations. Thus, overall, more than a thousand trawl hauls were made during the survey. This account is taken from Williams (1968).

Table 3 gives the catch rates in kg/hr^{-1} in one statistical area at the northern end of the

TABLE 3. CATCH RATES (kg hr^{-2}) BY DEPTH INTERVALS OF SOME FAMILIES OF FISH IN STATISTICAL AREA I
BETWEEN GAMBIA AND GUINEA

Family	Depth (m)							
	15/20	30	40	50	70/75	100	200	400/600
Squaliformes	47	18	64	26	33	61	89	11
Rajiformes	56	220	10	6	4	4	4	0
Ariidae	290	159	116	54	+	0	0	0
Carangidae	63	70	149	173	57	561	0	0
Clupeidae	74	24	29	5	43	3	0	0
Ephippidae	46	20	113	0	0	0	0	0
Lutjanidae	0	+	3	0	0	0	0	0
Mullidae	+	44	83	95	11	+	0	0
Pleuronectiformes	11	8	6	8	4	+	+	+
Polynemidae	126	147	104	0	0	0	0	0
Pomadasydae	157	466	176	52	3	253	0	0
Sciaenidae	96	122	50	15	32	1	10	0
Scombridae	+	0	+	29	31	47	+	0
Serranidae	+	4	11	10	10	17	3	0
Sparidae	16	216	159	139	141	362	+	0
Sphyraenidae	1	+	2	+	+	0	0	0

TABLE 4. ESTIMATED STANDING CROP IN
THOUSANDS OF METRIC TONS BY STATIS-
TICAL AREAS AND DEPTH ZONES

Statistical area	Depth (m)	
	15–50	50–200
1	298	108
2	188	40
3	18	27
4	7	18
5	11	11
6	14	30
7	27	38
8	13	12
9	35	12
10	21	20
11	13	26
12	25	49
13	79	79

survey area by families. The abundant families are the Carangidae, Pomadysidae, and the
Sparidae in 100 fm, but there are abundant stocks in shallower waters. Table 4 shows the
estimated standing crop in each statistical area in shallow water and in deeper water. The
greatest quantities were found in the northern end of the area, but a poor area was revealed
off Liberia and the Ivory Coast. A sharp day/night ratio was found for flatfish and gurnards.
A statistical study was made of the communities, each made up of an array of species; then
the environmental characteristics of each community were tabulated. In general it appeared

that the differences between communities were sharper than the environmental ones measured.

A considerable amount of data were collected on the nature of the seabed, partly from material sampled there and partly from the recorded damage to trawls. Such a rough ground survey is the first which has been explicitly executed. Later it will be shown that acoustic methods can be used to describe the surface of the seabed, but that "roughness" acoustically may have a different meaning to that used by a fisherman with a torn trawl. The systematic survey made in the Guinean Trawling Survey is the first step towards correlating such estimates.

The whole survey was executed in less than a year by three ships working continuously and systematically. Off the north east Pacific perhaps half as many trawl hauls were made over a period of a little more than a decade. No comparison need be made between either methods or results, but it is obvious that great areas can be explored fairly cheaply and quite successfully in rather short time periods. Earlier exploratory surveys, like those made in the North Sea by the International Council for the Exploration of the Sea in the first decade of the century, were much less successful than the Guinean Trawling Survey because they were organized less systematically. Thus, if a traditional survey can be executed as effectively, an acoustic survey supported by a trawl survey should be as effective and even more rapid.

8. *The positions of fisheries*

Fisheries are found in many types of regions; spawning areas, feeding areas, oceanic boundaries, and in upwelling zones. It should now be clear that the upwelling occurs near some oceanic boundaries and that the classification is to some extent arbitrary. Fisheries based on spawning grounds and seasons are limited to regular seasons, timed sometimes to weeks in limited areas in temperate seas. This is true of herring spawning on the bottom or of cod spawning in the coastal fjords; it is also true of tuna on a much larger scale, where spawning albacore are roughly limited to the North Equatorial current in the north Pacific. A feeding fishery is more complex, the fish crowding on to their food patches and dispersing from them. In the North Sea the dynamics of such processes are transitory and are perhaps not worth studying but in other waters as in the former Antarctic whale fisheries or in the upwelling areas, where food patches might be maintained for long periods, a study of such dynamics might be profitable. There are two types of boundary at which fisheries are found, where water masses meet and where water currents concentrate. The first is a water mass boundary as for example where the arctic cod are thrown up on the Svalbard shelf by the West Spitzbergen current and the cold water massed on the shelf opposes this movement and concentrates the fish into a fishery. The Kuroshio current is the western boundary to the north sub-tropical cyclone in the Pacific, where water transport and tuna are concentrated; thus deviations in the mass transport appear to cause displacement of the fishery. A more subtle form of boundary condition is that based on the herring fishery in the Baltic outflow. Food is produced earlier in spring than elsewhere in the North Sea because of the stable layer of Baltic water above the Atlantic water and where the food grows the herring aggregate.

The second physical base for a fishery is in the process of upwelling. The term includes true upwelling, divergence and some forms of advection (as for example across a dome). Dense fish populations are maintained in such regions by a continuous extension of the aggregation process in feeding and by the provision of very good survival conditions for the larvae. When we consider the true upwelling area, the surface waters are driven offshore by the wind and, to replace it, water is drawn from the depths to the surface. From the structure of current and counter-current generated in the upwelling system, it may be that the whole physical system can be treated as a biological unit. It is unlikely that such conditions apply to a small area like the Costa Rica Dome, but when one studies the populations off California and Oregon, the hake spawn in the south in spring when the upwelling starts and then move north presumably in the counter-current as the whole upwelling system shifts northward later in the season. It is in this sense that an upwelling system might be considered as a biological unit within which fish stocks are retained.

Fisheries, spawning or feeding, exist at regular positions on the migration circuits. So in place and in season there is a regularity to fisheries on which the fishermen rely. Fisheries at the oceanic boundaries, whether upwelling zones or the edges of ocean water masses, are special cases of the general rule. The migrations of some fish are quite well known; but for many they are unknown. In general a fish stock is one confined to a given circuit contained within particular current structures. Consequently the study of migration provides the fisheries biologist with the first step needed in his work of population analysis, the location of the stock.

CHAPTER 2

THE INDICATIONS OF FISH

1. *Introduction*

THE Suffolk liners who sailed to Iceland in the sixteenth century, and the Japanese long liners who sail all over the world today, searched for fish by catching them; today such traditional methods are being replaced by direct detection with echo sounders. In fact, the Japanese long liners do not use echo sounders yet because they depend on the sighting of birds. And in earlier times there were often surface indications of fish, which the fishermen learned to trust. Before the development of the direct detection of fish by acoustic methods is described in detail, a short account will be given of the older methods of finding fish independently of capture.

2. *Natural detection*

(a) *Seeing fish*

Fish that live on the bottom, the demersal fish, cannot be seen at all except on rare occasions; for example, sole and turbot have been seen swimming on the surface (de Veen, 1964). The pelagic fish can be seen more frequently. Herring have been seen at the surface, "making a ripple in the dark roughness", and on one occasion in the Bay of Cromarty in Scotland in 1780 a large shoal of herring appeared at the surface, driven by "vast numbers of whales and porpoises which beat the water into a foam for several miles" (Mitchell, 1864).

In certain fisheries, fish were sighted from a cliff top or a tall mast. The Cornish pilchard (*Sardina pilchardus* Walbaum) was caught by beach seines and the fishermen were warned of the approach of fish shoals by a "hewer" or lookout on the cliff top and they rigged their huge nets according to the direction indicated by a white ball on a pole (Wilcocks, 1882). In the Mediterranean, tall masts of 20 m or more in height (called vedette in Italy) were used for sighting tuna and signals were sometimes passed from mast to mast before they reached the fishermen. Strabo recorded the presence of such masts (Thompson, 1947) and they are still to be found in Turkey on the coasts of the Bosphorus and the Black Sea. Such methods depend upon the known movements of pelagic fish in bays, presumably in the direction of the currents.

Mitchell (1864) wrote that under certain atmospheric conditions, the glow of a herring shoal was enough to attract fishermen from a distance. The mackerel purse seiners off the eastern seaboard of the United States, in the seventies of the last century, searched till they found a "sight" of a good school near the surface. In the daytime, "the seiners prefer those schools which are 'cart wheeling' or 'going round and round in circles in a compact body in the act of feeding'." At night, "so remarkable is the phosphorescence thrown out from a large school of fish that it frequently seems to light up the surrounding darkness" (Brown Goode *et al.*, 1885). In Bigbury Bay, between Plymouth and Start Point on the south west coast of England, Mr. A. J. Pengelly has seen herring shoals as great phosphorescent heaps in the water; the fish were assembling on their spawning grounds, but at the actual time of spawning the shoals were no longer phosphorescent, or, in other words, were less active. Mr. Pengelly has also told me that he could distinguish shoals of mackerel (*Scomber scombrus* Linnaeus), herring, and pilchard by differences in their phosphorescent shapes in the water. Off the Californian coast, sardines were seen on moonless nights as the shoals glowed in the phosphorescent water. The purse-seine fishermen were able to distinguish sardines from anchovies or from jack mackerel by the pattern of their movements in the shining water at night. When the fish were close to the boat, a sharp bang on the gunwale startled them and they darted away (Scofield, 1929). The direct sighting of fish by phosphorescence depends on three factors; the fish must be quite close to the surface at night, there must be enough of the phosphorescent planktonic organisms, and the sea should be calm.

Sardine shoals have been located from a balloon off the southern coast of Brittany in France and they lay in long sausage-shaped shoals along the direction of the tide (Belloc, 1935). Breder (1951) watched aggregations of fish from the air in shallow water and could see the individual fish and these observations have laid the foundations of our knowledge of natural schooling behavior. Hardy (1924) flew over the East Anglian herring fishery in the autumn. He saw no herring, but he did see the patches of "gully water" which the fishermen themselves used as a sign of herring, because the fish were caught more readily in muddy water than in clear water. Because the sea is not very transparent, fish must be close to the surface to be sighted. The volume of the sea from which fish are caught is much larger than the thin layer of visibility near the surface, which in any case is only occupied at certain times. Most fish live deeper and when fish can be detected acoustically there is no longer any need for sighting.

However, fish have been sighted from the air. Blackburn (1956) demonstrated the presence of extensive shoals of pilchards off the coasts of Australia (*Sardinops neopilchardus* Steindachner) by means of aerial survey. Marr, in Cushing *et al.* (1952) describes the use of aircraft in surveying pelagic resources and in more recent years, light aircraft have been used to survey the menhaden off the eastern seaboard of the United States during a period of decline of the resource. But the most spectacular event recently in the field of direct detection is the use of satellites to find upwelling areas (Bowley *et al.*, 1969). What is detected is an area of very calm water which reflects sunlight directly to the camera on the satellite. Bowley has shown rather conclusively that such patches occur in the regions of divergence, like the eastern tropical Pacific Ocean, or above the Cromwell Undercurrent east of the Galapagos Islands. Of course, fish are not detected by this method, but the presence of very calm water may indicate a patch of upwelling to which fish may aggregate. The great advantage of satellite observations is that a very frequent time series can be elaborated.

(b) *Listening for fish*

In general, any sounds made by fish are lost in the noise at the surface, so if they are to be heard, it would be in calm and protected waters. It is not surprising that in the East Indies fish are caught by acoustic lures and that fishermen can listen to the sounds made by fish through the wooden hulls of their boats. They put their ears to the ends of their hardwood paddles and dip the blades in the water; indeed, some fishermen find the fish by sticking their heads underwater and listening to them. The black pomfret (*Stromateus niger* Bloch) is caught by an extraordinary method. A boy jumps overboard with a bamboo rod and cross bar and he floats with his arms over the cross, singing a low "ooh!"; when he starts to sing, the fish flock round him.

There are two other gears, the smaller and larger bamboo rafts and the smaller is called a rumpon. The rumpon is anchored to the bottom and is floated with palm leaves at the surface; the larger bamboo raft has an anchoring cable of up to 1000 m in length. As the leaves lift in the waves they rattle and, over a period of time, collect fish beneath them. In the Java Sea, 40,000–50,000 tons a year of carangids, clupeids, and mackerel are caught by rumpons and bamboo rafts.

This summary is taken from Westenberg (1953); he also suggests that the live bait method of catching tuna depends on the tuna hearing the bait striking the water because they can be attracted from a range of 35 m or more, beyond the normal visual range. The aggregation occurs very quickly, more so than could be explained by the diffusion of an odor. The waters in the East Indies are relatively enclosed and sheltered from the oceanic winds and the long deep swells. However, if carangids and clupeids can be aggregated by the crackle of palm leaves at the surface in a calm sea, presumably they could be aggregated in rougher seas by increasing the signal above the noise.

(c) *Fishermen's signs*

Because the only firm information on the distribution of fish comes from catches, fishermen have always been alert for any indications of fish in the environment by which they might predict their catches. Using a classification by appearance Bullen (1910–13) listed the six types of water which were used by the mackerel fishermen off Cornwall in England: 1. "Stinking water; dull leaden colour even in bright sunlight, so dense that a man looking over the side of a sailing drifter cannot see down to the keel." There is a smell of decaying seaweed characteristic of the normal algal outburst in the sea and mackerel are found in this water, and horse mackerel sometimes occur. 2. "Grey water"; has the same effect as the stinking water, but there is no smell and so presumably is an earlier stage of the same phenomenon. 3. "Blue or green water; this is the clearest water and holds the prospect of a fair catch." 4. "Yellow water; there is a yellow tint and is rather dense when viewed either in sunlight or under a dull sky; often it appears in patches and teems with minute animal life." The highest catches are made in such waters but light catches are also made there. 5. "Oily appearance in spates (patches); this is a sure indication of drift net fish." 6. "Milky appearance inshore; small mackerel are sometimes found in the turbid patches."

The fishermen's signs in the water may be grouped under three heads, differences in the plankton, the presence or absence of oil, and the turbidity of the water. In the North Sea herring fishery, many drifter skippers had the propeller and rudder painted white, so that the density of "feed" (animal plankton) could be roughly estimated against the white background in clear water. In the East Anglian herring fishery, as noted above, the driftermen looked for "puddly" water or "gully" water, which are patches of mud and sand stirred up from the banks. Fish were caught by drift nets more readily in the "puddly" patches than in the "sheer" or clear water. Another sign in the North Sea herring fisheries is the presence of "ginger beer" water, when the bubbles, rising at the side of the vessel under way, rise more quickly than usual.

A more general sign is the presence or absence of predators. Gulls and gannets, porpoises and whales gather round herring shoals, like terns on sprat shoals, and birds or whales can be seen from some distance. In the North Sea herring fisheries, the presence of gannets was regarded as a very good sign, particularly if there were many of them. This is also true in the South African pilchard fishery and the Peruvian anchoveta fishery where gannets and pelicans indicate the presence of fish. In the great oceanic fisheries for tuna in the Atlantic and Pacific the fishermen make use of the presence of birds to indicate the tuna. A good fisherman can see them up to three miles away, further than sonar range underwater. In these fisheries a good lookout is still more valuable than an echo sounder.

In the North Atlantic the sign used most frequently by the trawlermen is the presence of other trawlermen, irrespective of nationality. Very often the trawlers gather in "arctic cities" where the fish concentrate. Not only are the "cities" visible, the lights of the ships looming on the low clouds, but they can be detected by radar at a fair range. Further, one of the main jobs of the radio operator is to eavesdrop on the continuous radio chatter and to determine the position of successful rivals with the direction finder.

(d) *The end of natural detection*

Fish can be seen in the water under special circumstances and in the East Indies they can be heard, but because the sea is opaque and the surface is noisy, the majority of all fish is neither seen nor heard. The signs of fish indicate their presence indirectly, but amongst the pelagic fish, at least, fish are now detected directly with echo sounders by the fleets of most countries. There are some fish, like plaice, which cannot readily be detected acoustically and so plaice trawlermen search by catches from haul to haul. Cod are readily detected and many cod trawlermen use echo sounders. So the use of many of the fishermen's signs has tended to disappear.

3. *Ultrasonic detection*

(a) *The echo sounder*

A transducer of nickel, barium titanate, or lead zirconate converts electrical energy into acoustic energy at a fairly narrow band of frequency in short pulses of about 1 msec. It is mounted on the bottom of the ship and when the ultrasonic pulse is emitted from the transducer it is constrained into a fairly wide cone towards the seabed. The pulse travels towards the bottom where it is reflected back to the transducer; the true depth is the shortest path between transducer and seabed, so the first signal to return in echo is the best estimate of depth. Subsequent signals come from longer path lengths at an angle to the axis of the sound beam. Consequently the signal from the seabed appears to last on the record for several milliseconds. Indeed the harder the bottom, the longer the apparent seabed signal as it is returned from a wider angle with a longer path length. The speed of sound in sea water is 1500 m sec^{-1} (approximately) and so half the time interval between transmission and the reception of the first returned signal, multiplied by the speed of sound, is a measurement of depth in meters.

When the signal from the bottom has been returned, a new transmission can be initiated. The pulse repetition frequency varies inversely with depth and there is one echo sounder at least in which this characteristic is exploited to the full. But, second or further multiple echoes from the bottom can be recorded apparently in mid-water. Scientists may accept such failures in order to exploit the pulse repetition frequencies to the full for particular purposes, but echo-sounder manufacturers must restrict the pulse repetition rates to fixed values for certain depth ranges or phases; they cannot afford to record the seabed at apparent depths less than the true depth; of course, multiple echoes of the seabed occur at multiples of the true depth. Obviously the pulse repetition rate is a function of the maximum depth chosen for a given phase. If repeated as quickly as possible for the depth, the sequence of transmissions yields a quasi-continuous record of the depth of the sea. The signal from the seabed is measured in mV (and that from fish in mcV) so that it has to be amplified for display on a paper record or on a cathode ray tube (C.R.T.). The paper used is either a wet starch iodide paper of fair dynamic range or a dry paper (of various forms) of rather less dynamic range. The C.R.T. is used in an A-scan form and depth is indicated on the time base; signals from fish and seabed are shown as deflections from it. The C.R.T. has the advantage that the received voltages are accurately recorded, but the disadvantage that there is no permanent record or "memory". The paper record is semi-permanent and so has the quality of "memory" but voltages are only recorded accurately within the fair dynamic range of the paper. Consequently, most modern echo sounders are equipped both with a C.R.T. display and a paper recorder.

It is not my purpose here to describe the circuits used to operate an echo sounder. However, there are three items of equipment which should be noted: (a) a power pack which generates pulses at the required frequency (the electrical peak power used in fish finding echo sounders ranges from 1–8 kW); (b) a triggering mechanism which sets the power transmission in train subject to the restraints in pulse repetition frequency noted above; and (c) an amplifier which raises the received voltage sufficiently for display. Subsequently, under the heading of calibration, the power pack and amplifier will be referred to in some detail; the electrical power (and/or acoustic power) must be standardized and the amplifier must be checked for its rated characteristics if the echo sounder is to be used in a quantitative manner.

It is likely that the first signals received from fish using an ultrasonic transmission were noticed by Rallier du Baty (Hodgson and Fridriksson, 1955) on the Grand Banks in 1926. The recording echo sounder was invented by Wood *et al.* (1935) of the Admiralty Research Laboratory of the Royal Navy, in Britain; the invention was developed by the firm of Henry Hughes in Britain. It was first used on a fishing vessel, the steam trawler Glen Kidston, on a voyage from Hull to Bergen in 1933. The same type of machine was used by Sund (1935a, b) to make his classical record of cod in the Vestfjord, which showed the layer of fish quite clearly (Fig. 25). At about the same time, fish were recognized with an echo sounder by Edgell (1935) on board H.M. Survey Ship Challenger in April 1933. Off Start Point a mountainous echo appeared which obscured the bottom signal and rose to within a few fathoms of the surface. The depth of water was checked with a leadline and it corresponded with that recorded on the chart. This mountainous echo was probably received from fish (Fig. 26); twenty years later, at the same time of the year, and in the same place, large echo traces were identified as pilchards (Cushing and Richardson, 1956). During the same period (1933–7) the late Skipper Ronnie Balls (1946) was working with a neon tube receiver (a Marconi 424). The bottom echo appeared as a peak on a display which resembled a C.R.T. and fish echoes appeared as single or multiple peaks in the mid-water and Skipper Balls found that he could associate those in mid-water with the presence of fish. One of his drawings shows a record on the echo sounder from a shoal of herring, part of which was breaking surface alongside his drifter.

Only the essentials of the simplest recording echo sounder have been described, the type of machine which was used by the pioneers of acoustic fish detection nearly forty years ago. Since that time many varieties of echo sounder have been built for many purposes, some of which will be described in later pages. However, it is not part of our purpose to follow the minor adaptations to the different needs which fishermen may have; for example, some recording echo sounders are designed to use the least quantity of paper and save money,

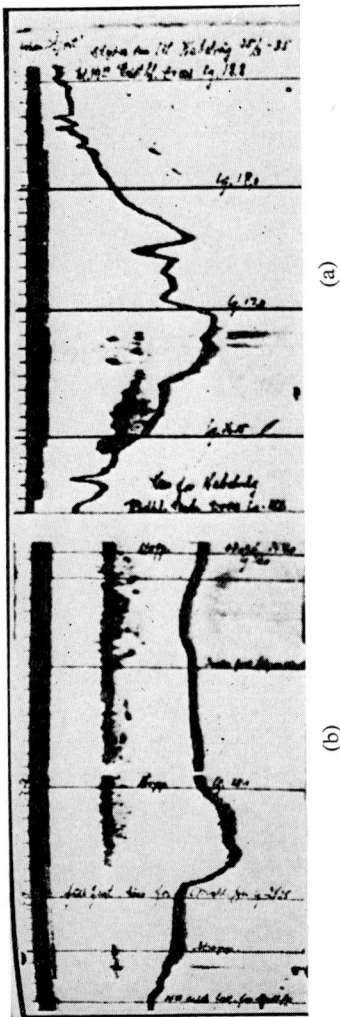

(a)

(b)

Fig. 25. A trace of cod recorded in the Vestfjord (Sund, 1935); the section (a) across the fjord shows the spawning shoals on its northern edge and that along the fjord (b) shows the constant depth of the layer.

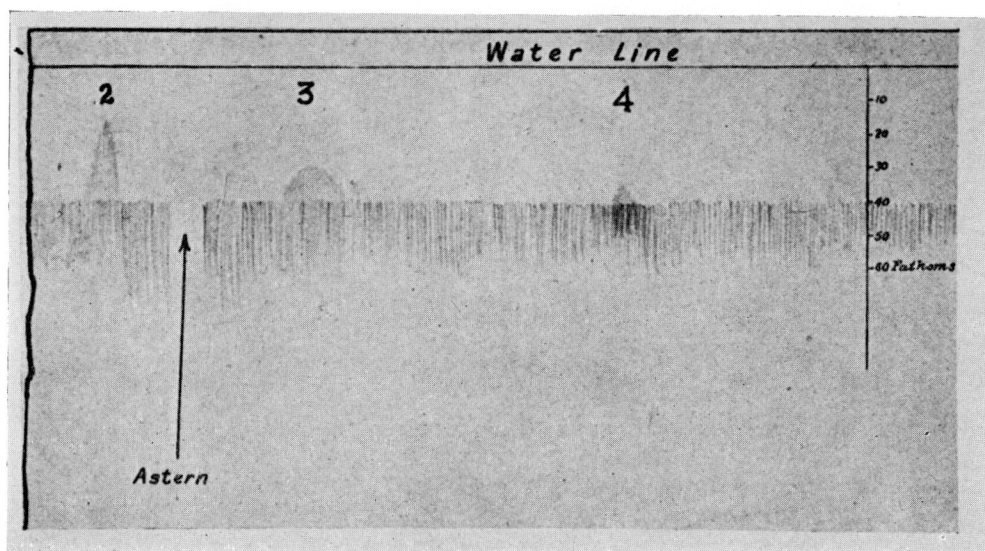

FIG. 26. The mountainous trace from fish (possibly pilchards) recorded by H.M.S. Challenger in 1933 off Start Point in the English Channel (Edgell, 1935); (reproduced by kind permission of the International Hydrographic Bureau).

and of course, lose information which the fishermen have paid for. So the echo sounder will be treated as if it consisted of a transducer and a display with little reference to the essential intermediate electronics or to trivial modifications in display.

(b) Description of fish traces

The paper record from an echo sounder appears to describe a section of the sea along the ship's track. In so far as it does do so, features on either side of the track are not recorded. If fish shoals were always spherical or spread in flat layers, the transect would sample them properly; but fish sometimes lie in long sausage-shaped shoals on the axis of tide or current and under these circumstances they are improperly sampled. As will be shown later the shapes of fish shoals as described by the ARL scanner are diverse and they vary under different conditions.

The echo sounder measures the true depth of the sea correctly, but the depth of any object in mid-water, except a layer, is nearly always over-estimated. The depth of the layer is correctly measured for the same reason that the depth of the sea is measured properly. Consider a fish or a fish shoal swimming at constant depth on a diameter track across the cone of the sound beam. Its range decreases from its point of entry into the beam until it reaches the axis of the beam and from there its range increases to the point of exit. Such a track would appear on the paper record as the hyperbolic "comet trace" (Figs. 27, 28 and 29) and for a fish or fish shoal on a diameter track, the peak of the hyperbola is a measurement of its true depth. However, such a track is a very uncommon one and on any other track at the same depth, a chord track, the hyperbola is shorter and its peak does not represent a true depth, but a true range, which is greater than the depth. The length of the hyperbolic trace depends upon the distance of the chord track from the axis at its nearest point.

The way in which traces are generated by single fish and by shoals smaller than the sound cone is illustrated in Fig. 27. The differences between a single fish, a spherical shoal, and a lozenge-shaped shoal are shown as differences in vertical extent; the trace from the lozenge-

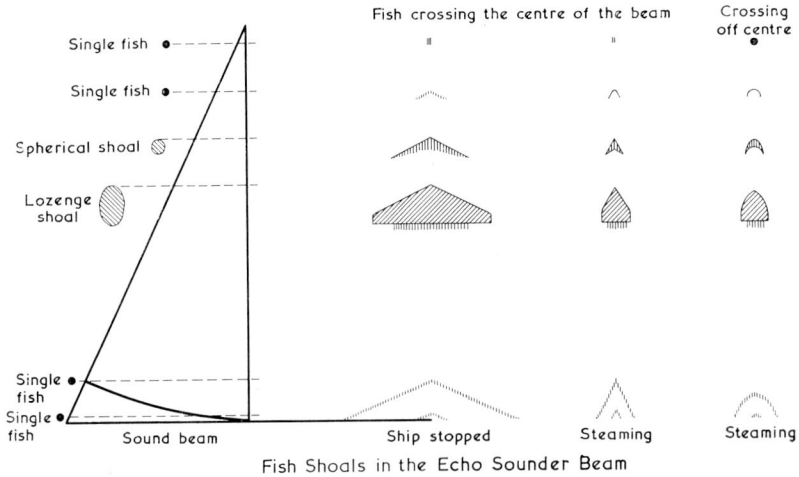

FIG. 27. How fish traces are generated; on the left are shown tracks of fish or shoal across the sound cone. On the right are shown the traces, with ship stopped and steaming (Cushing, 1963).

shaped shoal is shown with an extension as if sound was reverberating between fish. Evidence for this assertion will be presented below. Further differences in both chord tracks and diameter tracks are given, as are the differences between a steaming ship and a stopped one. The spherical wave strikes the bottom on the axis which means that there is dead ground in which the fish signals can be masked at an angle to the axis by the bottom signal received on the axis.

A layer of single fish at one depth would appear as a distribution of hyperbolae, a few very long ones near the diameter track, most of intermediate length, and many very short ones near the edge of the beam. Thus the hyperbolic trace pattern is an effect of examining a fish swimming at a constant depth with a spherical wave. Consequently, the depths of nearly all single fish and shoals smaller than the sound beam must be over-estimated; single fish can be detected at their true depth transiently but at maximum range.

For normal echo sounders with an apex angle to the sound cone of 40° (or half angle from the axis of 20°), the over-estimate of depth can amount to 10% in extreme, but is usually about 5%. Most fishermen and fisheries biologists are interested in depth, which for single fish and small shoals is indeterminate; therefore a statistical treatment is needed which will be given later. The hyperbola is described formally as follows:

$$r = (z^2 + d^2 + s^2)^{\frac{1}{2}} \qquad (1)$$

where r is the range from the transducer; z is the true depth of the target; d is half the length

FIG. 28. A series of "hyperbolic" traces recorded at a maximum range of 20 fm and a minimum range of 4 fm with one exception which is approximately the depth of the targets (Cushing, 1963).

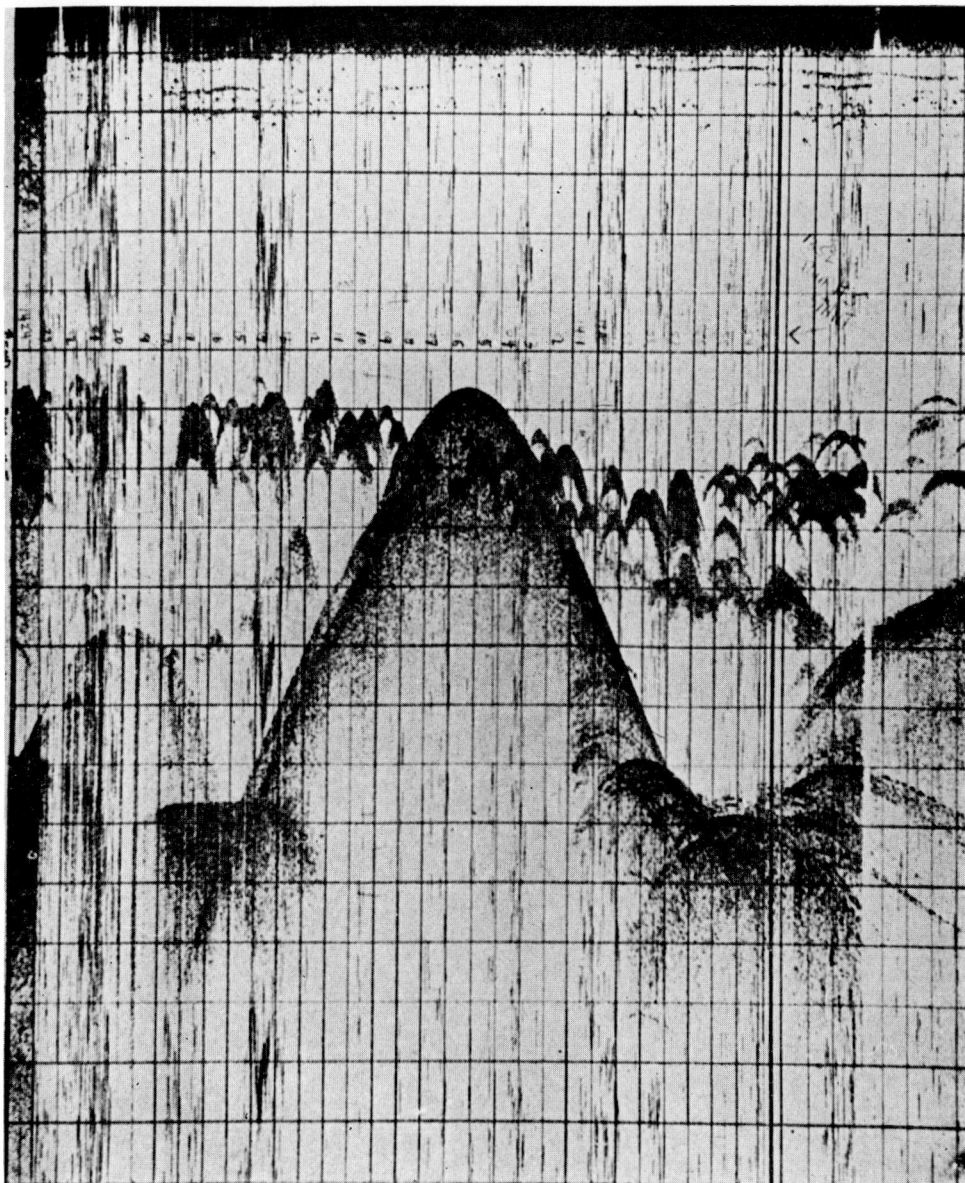

Fig. 29. A series of striking "hyperbolic" traces from the Deep Scattering Layer at a depth of 70 fm in the Atlantic; the large trace is one from the seabed at a depth of 940 fm, on a different phase. Each trace is of a small shoal, the estimated diameter of which is given by the vertical extent of the trace from its apex, 3–15 fm; the depth scale is in units of 10 fm (Hoffman, 1957); (reproduced by kind permission of the International Hydrographic Bureau).

of the chord track; and s is the distance from the point of minimum range to the axis of the sound beam.

At different ranges or at different speeds of ship relative to the target, the shape of the hyperbola changes on the paper record, flattening with range and sharpening with increased relative speed of ship to target. With increased range, the average track lies further from the axis of the sound cone, so the hyperbolic trace endures for longer. At a greater range, short of the maximum range for a particular size of fish where it is recorded transiently on the axis only, the effective sound cone decreases in size and the hyperbolic traces are shorter. However, extensive hyperbolic traces have been recorded; Fig. 28 shows traces of targets detected at ranges of about 20 fm in an apparent depth of about 4 fm. So the transducer, with a half angle of 24°, appeared to be a very wide angle receiver, which it is, if all the side lobes received the signals. But the target strength, or acoustic size of the fish was very high. Valdez (1961) has recorded such an exaggerated hyperbolic trace from a dolphin, which was identified by the animal's ultrasonic signals scattered like noise in bursts across the paper record; the signals are scattered in such a way because they are received independently of the timing of the echo-sounder's transmission. Hence it is possible that the remarkable signals recorded, as shown in Fig. 28, were received from porpoises or dolphins hunting herring off Dunmore in Southern Ireland (where the drifter making this record was fishing).

Figure 29 shows a series of hyperbolic traces from the Deep Scattering Layer in the Atlantic; the layer is an oceanic layer, comprised probably of fish, and it is described more fully later. On the same record is shown a series of bottom traces on a deeper phase, which means that the bottom is far below the scattering layer. The layer consists of a number of small shoals shown as hyperbolic traces; an estimate of the size of a shoal is given by the vertical extent of the trace from its apex. The diameters ranged from 3–15 fm. The bottom traces show the same hyperbolic character as the fish traces for the same reason, that a bottom feature changes its range to the transducer as the ship passes over it. As compared with fish, the bottom signals may be three orders of magnitude greater in amplitude and very large areas of the seabed were detected by the spherical wave on the occasion that Fig. 29 was recorded.

The simplest echo trace is that of a single fish crossing the sound beam at constant depth. Not only is it of the expected hyperbolic shape, but it is also of constant vertical extent, of one pulse length only. The pulse length is the duration of the sound pulse in time, which ranges from one tenth to two or three milliseconds; for most echo sounders there is a choice between 0.25, 0.50 and 1.00 msec. The pulse length is measured in time and at a sound speed of 1500 m sec^{-1}, a pulse length of 1 msec = 1.50 m in transmission; in echo, the length is halved on the display because the time to return a signal to the transducer is twice what it would take to reach a target, and so the pulse lengths of 0.25, 0.50 and 1.00 msec correspond to distances of 0.188, 0.375 and 0.75 m on the display. Figure 30 shows such a trace from a single fish, a "fingernail" trace, which is recorded for one pulse length only, throughout successive transmissions; because the character is so distinctive, single fish can be easily recognized. There is another trace which extends vertically for only one pulse length and it is recorded from a very thin surface, the material in suspension at the density discontinuity in a thermocline; Fig. 31 shows a trace from such a discontinuity (which is discussed further below) and it is composed of signals received from the axis only, which are not strong enough to be received from any angular path (Cushing et al., 1955). Hence, the trace records the true depth of the discontinuity for the same reason that the depth of the sea itself is recorded correctly.

A small shoal of three or four fish, quite close together, but at slightly different ranges, yields a trace thicker than one pulse length because the pulses overlap in time at the receiver as a consequence of the small differences in range. So the echo trace from a small shoal is a hyperbolic one which is thickened at the center (Fig. 27). It is a very common trace indeed

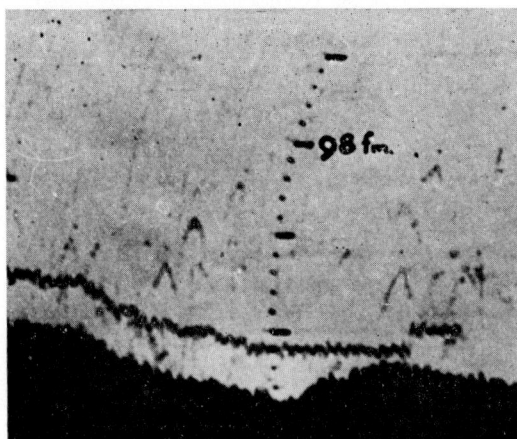

FIG. 30. A "fingernail" trace (Cushing, 1963).

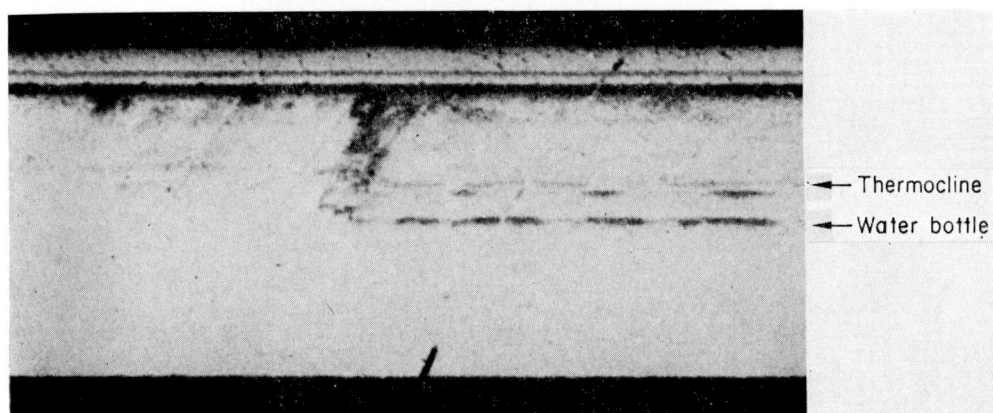

FIG. 31. An echo trace from a thermocline in Windermere in northern England; the thermal discontinuity was $1.5°C\ m^{-1}$. The trace was denser when animals were present and the thermal discontinuities were less sharp (Cushing *et al.*, 1955). The depth of the thermocline is marked by the upper arrow. The lower arrow indicates the signal from the frogmen.

and it is likely that the vertical extent is a true estimate of shoal diameter, because the chance of reverberation between a few fish is rather low.

A larger shoal is represented by a "plume" trace, which is given by a spherical shoal or a lozenge-shaped shoal (an ellipsoid with its major axis vertical). The head of the plume is effectively a hyperbolic trace, so the shoal causing it is one smaller than the sound cone. Furnestin (1953) published some traces of *Sardinella aurita* Val. off the coast of Morocco which were much longer vertically than they were horizontally. The true shape of a fish shoal is distorted on most common displays because the vertical scale is much exaggerated in relation to the horizontal scale.

There are large traces from big shoals. Some of these are so intense that the trace extends, at a point below the center of the trace, in a "hangover". Such a trace, if close to the bottom, sometimes extends far beyond the furthest signal from the bottom itself. An explanation is that sound energy reverberates between fish or from fish to bottom and returns to the

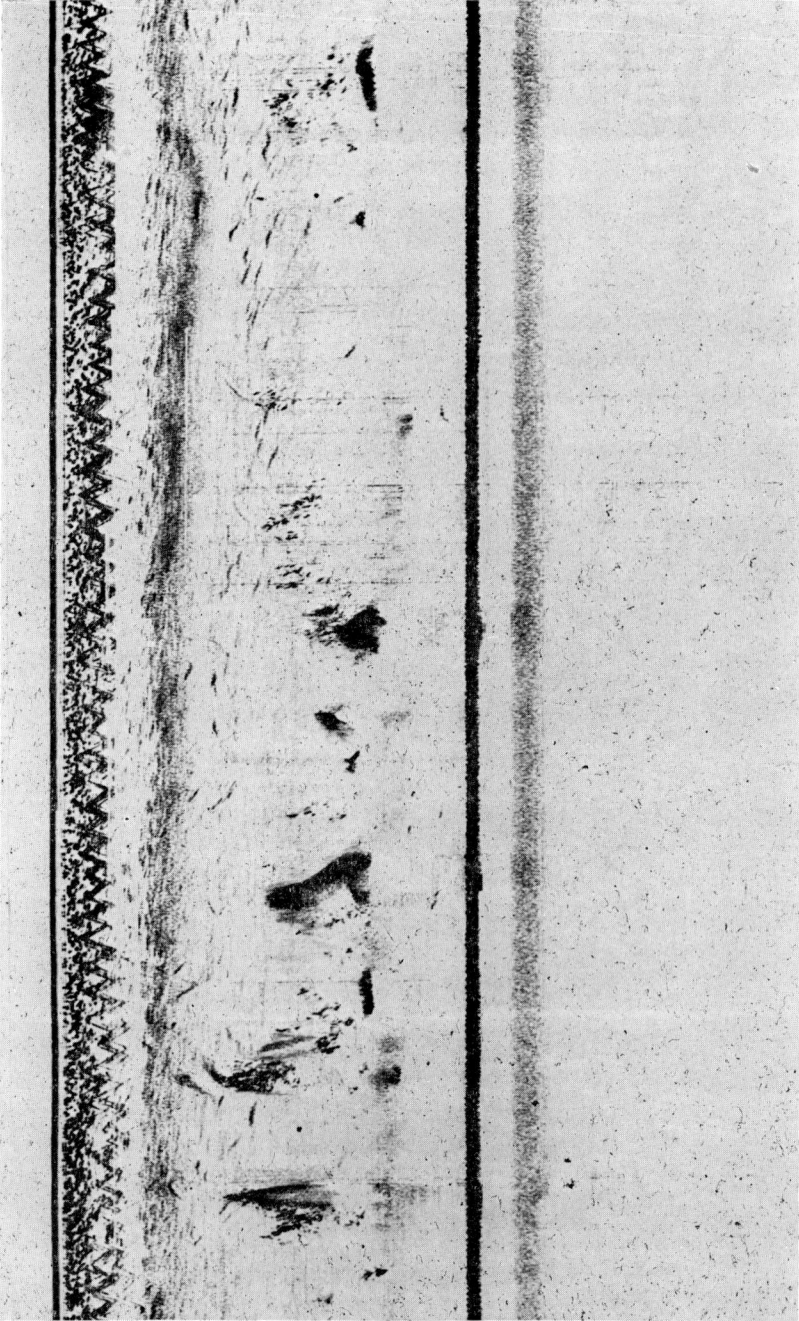

FIG. 32. Large traces of pelagic fish recorded by the Lowestoft 100 kHz echo sounder; the shoals were resolved into single fish to some degree, but reverberation is also shown, particularly on the left hand shoal. There are three layers of single fish in the middle one; the discrimination into single fish is not sustained. The cause of the saw tooth pattern is unknown.

transducer much later than the direct echoes. Figure 37b shows clear examples of the extension of the bottom trace beneath a fish trace; Fig. 32 shows a remarkable trace recorded by the Lowestoft 100 kHz echo sounder. The signals from closely packed individuals are distinguished from the smoothed signals of reverberation which returned after the last signal from the shoal itself in the center of the figure. Thus, if fish are packed together closely enough, sound can reverberate for a considerable time. In a later chapter, when the quantitative treatment of voltages received from shoals is considered, reverberation constitutes a source of unwanted signals; but the clear separation of signal from reverberation shown in Fig. 32 might provide a mechanism for rejecting the unwanted reverberation.

FIG. 33. A trace from the deep water south west of Hope Island in the Barents Sea in late summer; the larger traces in shallow water are probably from capelin and the hyperbolic traces in 60–80 fm are probably from larger fish like cod (Cushing, 1963).

The shapes of traces are diverse, the variety being caused by the size of shoal in relation to the sound beam, the depth of the shoal, the ship's speed, and the speed of the recorder. Figure 33 shows a trace from the deep water south west of Hope Island in the Barents Sea in late summer. The massive traces in the shallow water are probably from large shoals of small fish like capelin (*Mallotus villosus* Müller) and the small "crescent" shoals in the deep water (in 60–80 fm) are probably from larger fish like cod. It is a general rule that within the range of light penetration bigger animals in the sea live in greater depths than the smaller ones; again smaller animals tend to be much more abundant than the larger ones and such differences in abundance are shown as differences in density. Hence the characters of the two traces are to be expected on general grounds. Figure 34 shows hyperbolic traces of *Pseudosciaena manchurica* Jordan and Thompson in the East China Sea. The tips of the hyperbolic traces appear to sink into the bottom. The depth of the bottom is correctly estimated, but the depths of the fish traces are over-estimated. Because the sound strikes

FIG. 34. Traces of *Pseudosciaena manchurica* recorded in the East China Sea; it will be seen that the points of the hyperbolic traces disappear into the bottom trace (Cushing, 1963).

the bottom in a spherical wave there are patches of "dead ground" off center in which fish can swim and not be recorded as shown diagrammatically in Fig. 27. In deep water fishing, full hyperbolic traces tend to represent fish above the headline of the trawl. Catchable fish are presented as small traces merging with or "hard down" on the bottom which are the truncated hyperbolae shown in Fig. 27. The signal from fish is distinguished from the bottom signal by its magnitude, being about 1/100 to 1/1000 of the signal from the bottom. On the C.R.T. display, the writing speed across the screen of the weaker signal from fish is less than that of the stronger signal from the seabed. Consequently the signal from fish is brighter than that from the bottom and the weak fish signals are clearly distinguished from the strong and contiguous bottom signals. Such signals are necessarily from fish very close to the bottom and are therefore catchable. An analogous effect can be achieved for the paper recorder by using "white line", "black line", or "gray line" devices which obliterate most of the bottom signal and leave only the very thin line of the first signals in time. Against this thin line, the fish traces stand out and so are distinguishable from the trace of the seabed (Fig. 35). The white, black, or gray line is achieved by the strong bottom signal effectively switching itself off about 1 msec after it has first been received. Fish signals are separated from bottom signals on the display but those in the "dead ground" cannot be detected with

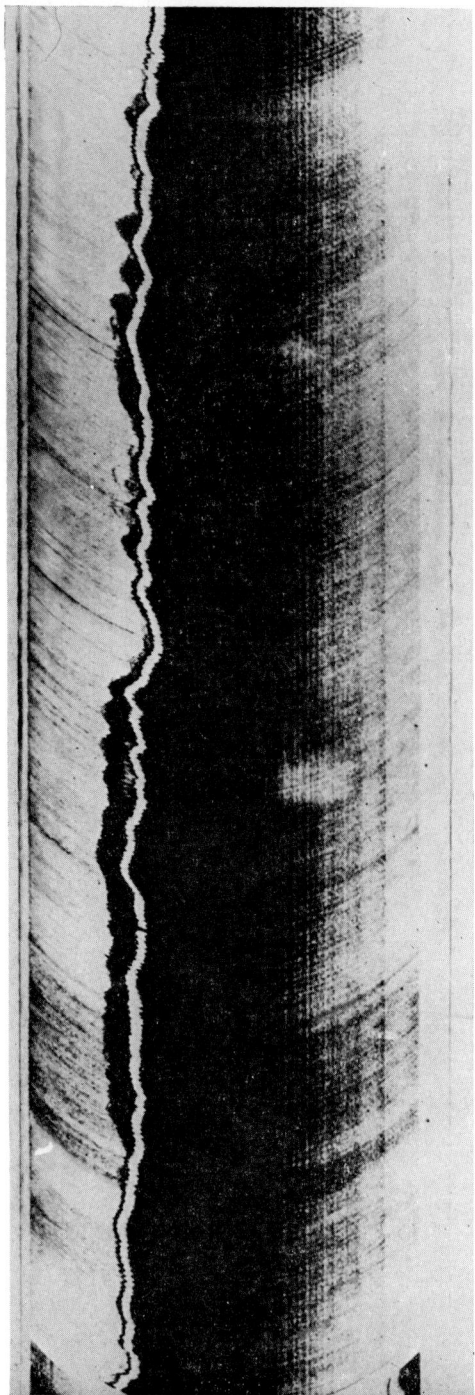

FIG. 35. An echo record with the "white line" technique; the thin trace of the bottom allows the fish signals to be distinguished from the seabed signals (with kind permission of Kelvin Hughes Ltd.).

this device. This problem can be mitigated by using an echo sounder with a beam so narrow that the area of dead ground is reduced. Whatever the beam angle, the pulse is transmitted in a spherical wave. Consequently there must be a dead ground in any case, but with a narrower beam its area is reduced, and so is the average height of the dead ground, and so the chance of detecting a fish in the dead ground is increased. However, the sampling power of a narrow sounder is reduced and the density of fish is quite independent of the nature of the echo sounder and therefore it might be necessary to increase the sampling power in some way; ideally the sampling power could be augmented by scanning a conventionally wide beam.

There are only two other forms of trace, a continuous layer and a mountainous trace. A mountainous trace requires little description: it is just very large. Frequently the layer can be analyzed into a number of hyperbolic traces of different sizes like those of the deep scattering layer as in Fig. 29. But in deep water, sometimes, thick layers have been found which cannot be analyzed into their component shoals. However, a more intense signal is sometimes received from the middle of the layer and not from the top or bottom; it is possible that it represents the signals from a middle angle where they are expected to be more abundant.

(c) *Fish traces from different species*

The shape of a fish trace on the paper record is governed by the size of shoal, speed of ship, and range. But fish shoals of different species vary in size seasonally. For example feeding shoals are small and rather active, whereas spawning shoals may be large and inactive. Further, the animals live at different depths at different seasons; the arctic cod live at about 50–70 m in the Vestfjord in Northern Norway, where they spawn, but at 200–300 m on the edge of the Continental Shelf off Bear Island, where they feed. Hence, two of the major variables governing the shape of fish trace change considerably during a season. It might be expected that the variation within a fish species would be as great as that between species.

Figure 36 shows a set of echo traces of herring, which is as diverse a set of echo records as can be found. Figure 36a is a trace of single fish in the Skagerrak, between Norway and Denmark; in Fig. 36b, part of the extensive (up to 27 km in length) and rather diffuse layer of spawning herring, which used to be found at night near the Sandettié Light Vessel in the Straits of Dover, is shown on an expanded scale to comprise many small and discrete shoals. Figure 36c illustrates a layer of Atlanto-Scandian herring in 138 m in the Norwegian Sea as the fish approach the Norwegian coast to spawn in winter. The plume trace (Fig. 36d) is often found in a feeding fishery in the North Sea and the mountainous or tower trace (Fig. 36e) is usually seen on herring spawning grounds both in the daytime and at night. This series of records can be arranged in a seasonal sequence; the spent herring are scattered (Fig. 36a), the feeding fish gather in small shoals (Fig. 36d), the pre-spawning fish in deep water live in a continuous layer spread across a considerable area (Fig. 36b), and when the fish spawn they appear to be firmly in contact with the bottom (Fig. 36e), which is not surprising because they lay their eggs there. These traces were collected and published by Cushing (1963). The herring has been identified as taking up all the forms of echo trace possible and this is a character which they share with other fish species.

Three echo traces of pilchards show the same sort of variety; Fig. 37a is a trace of the small feeding shoals in early summer off Start Point in Devon in S.W. England. Figure 37b shows "plume" traces off Cornwall, in S.W. England, indistinguishable from "plumes" of

FIG. 36b.

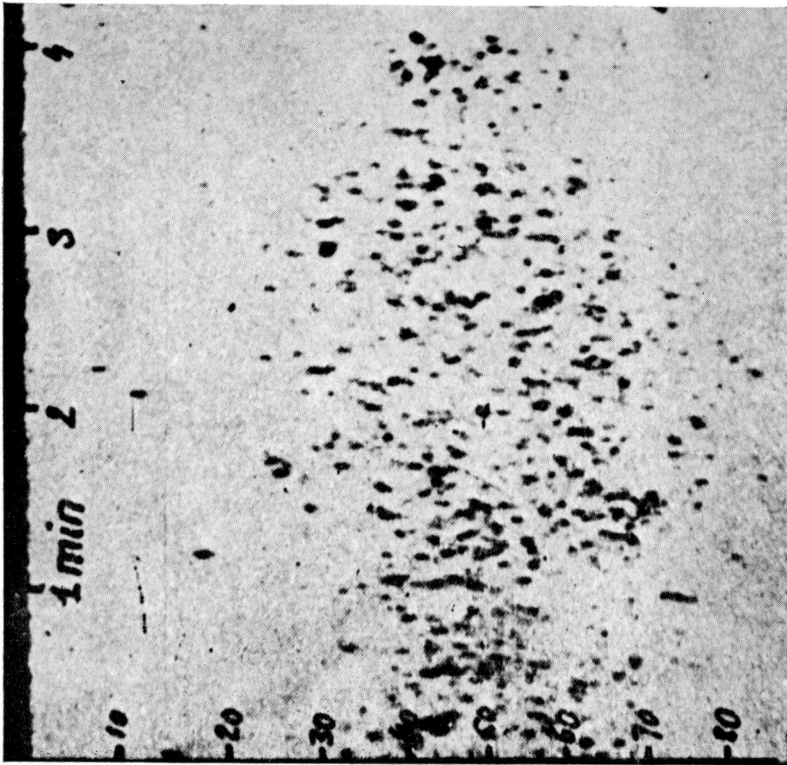

FIG. 36a.

FIG. 36. a, A spotty echo trace of herring in the Skagerrak in the north east North Sea. b, An echo trace of an extensive and diffuse layer of herring near the Sandettié L. V. (Cushing, 1963).

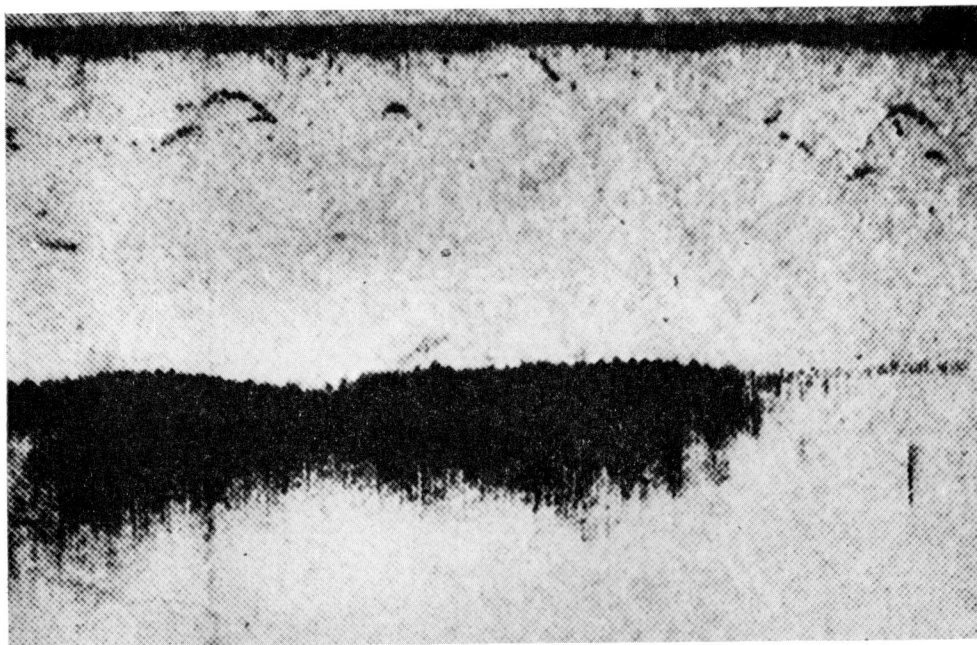

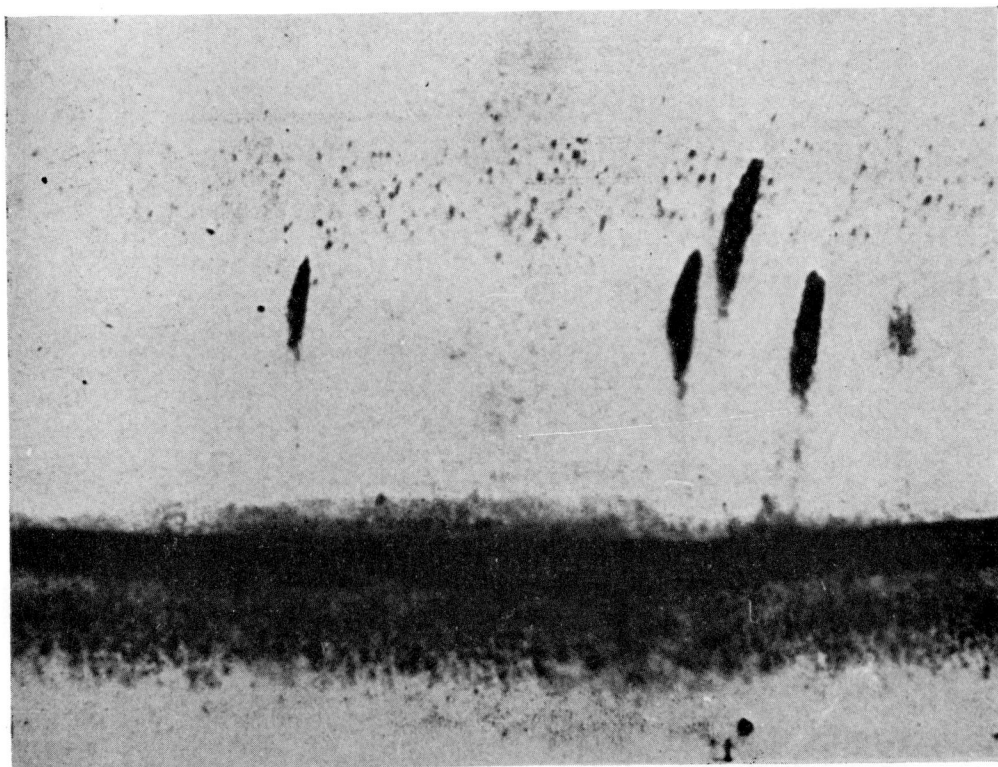

Fɪɢ. 36. c, A "veil" echo trace of herring at 150 m in the Norwegian Sea. d, The plume echo traces of herring in the feeding fisheries (Cushing, 1968).

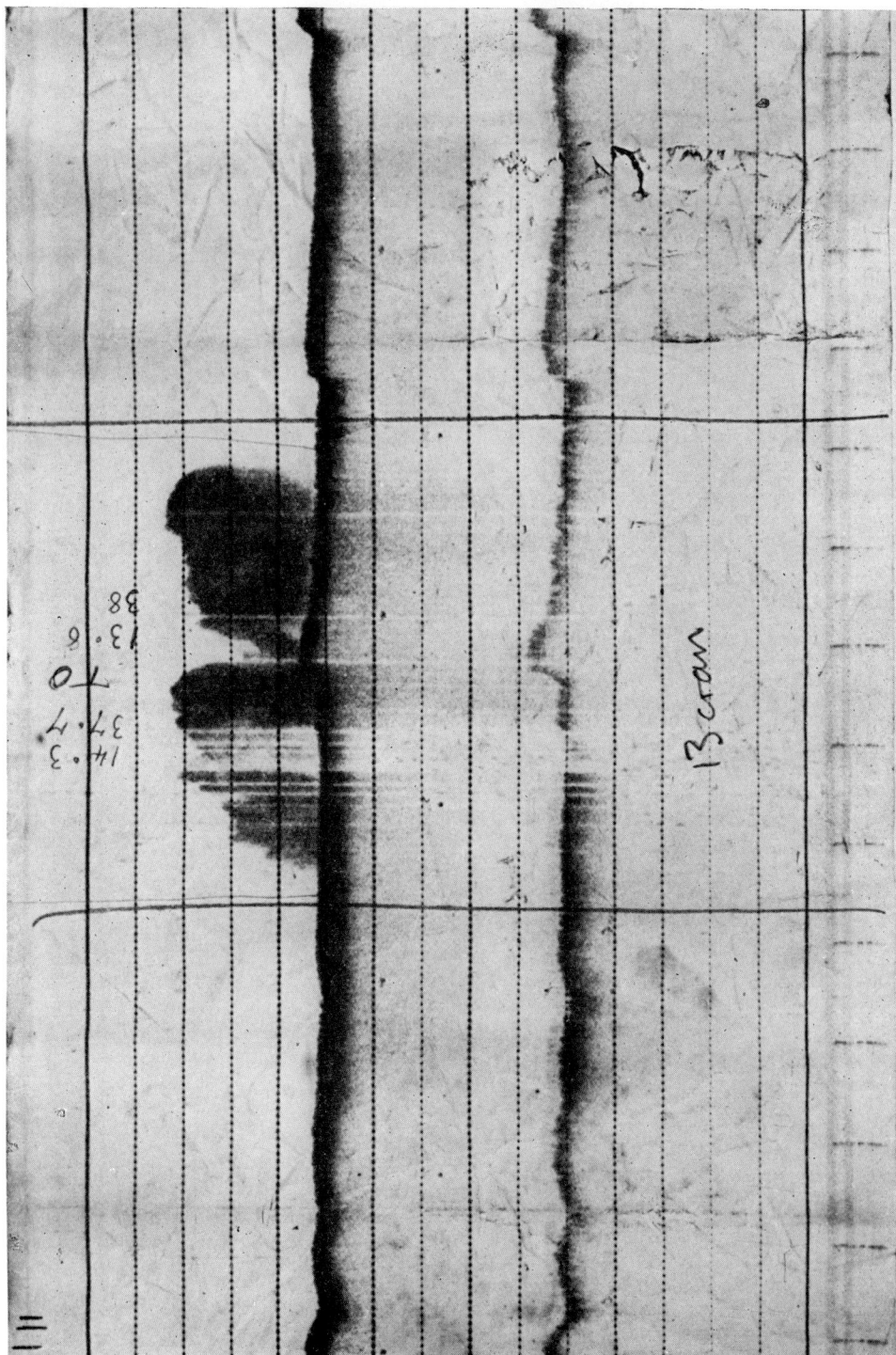

FIG. 36. e, Mountainous echo traces of herring often found in spawning fisheries (Cushing, 1963).

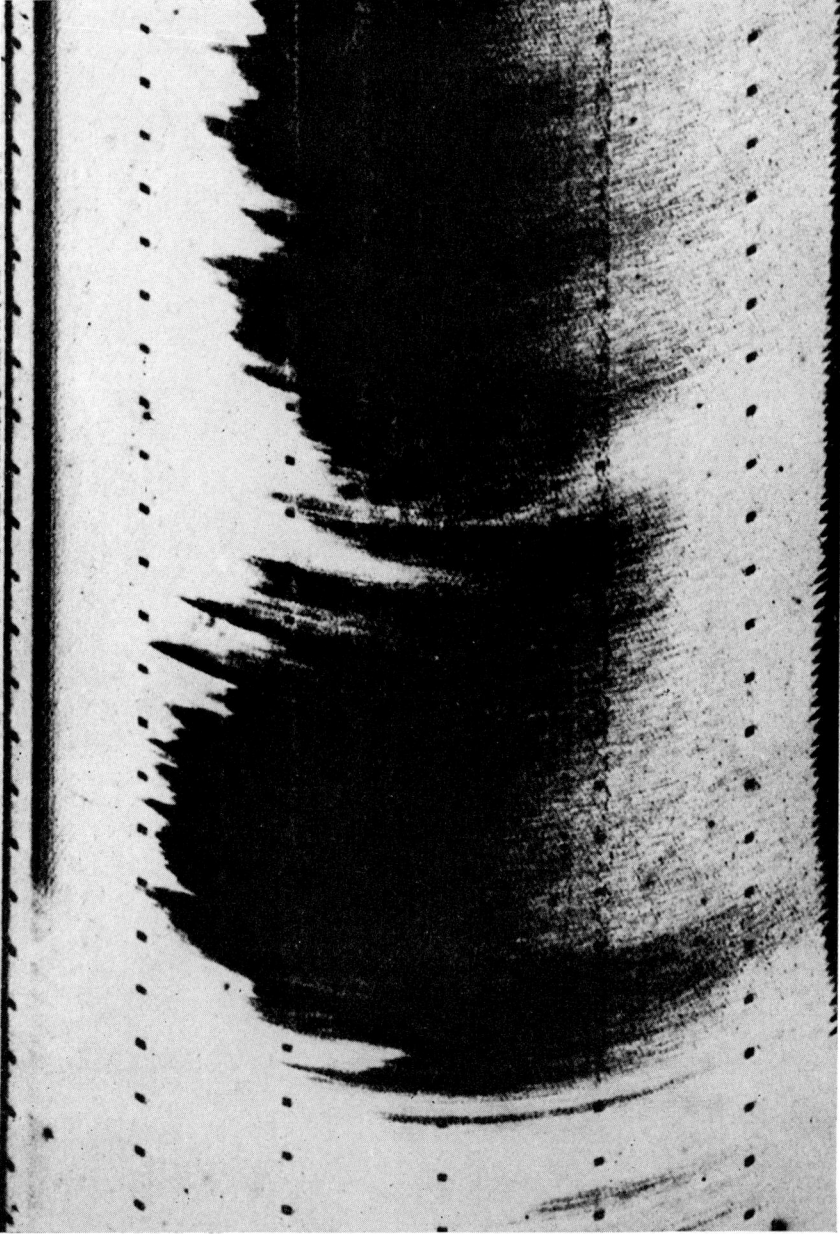

FIG. 37a. An echo trace of small feeding shoals of pilchards in early summer off Start Point, in the English Channel (Cushing, 1963).

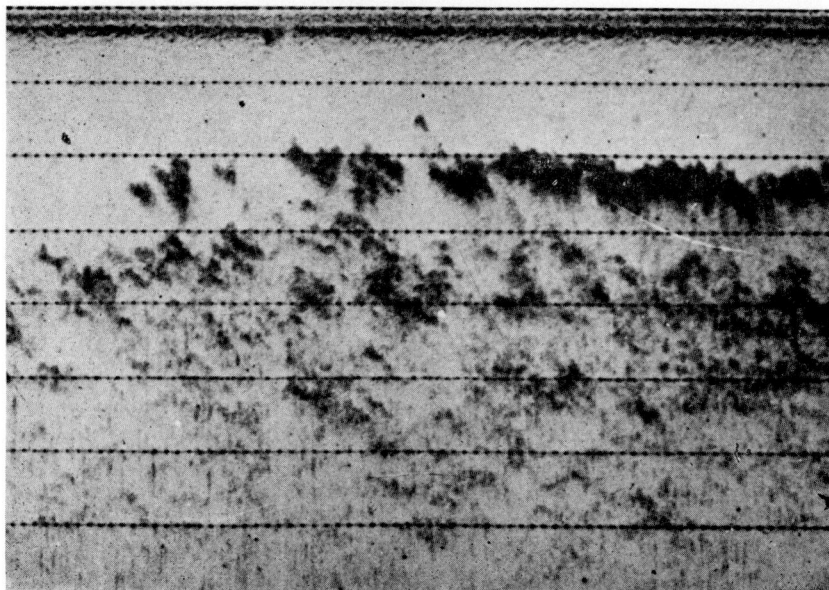

FIG. 37b. Plume echo traces of pilchards off Cornwall (noted the reverberation extending below the bottom trace from the left hand plumes) (Cushing, 1963).

herring. Lastly, in Fig. 37c is shown a layer of pilchards south east of the Dodman Point in Cornwall where they winter. Traces of cod do tend to have a different character from those of herring. Figure 38a is a trace of small shoals of cod in 100 fm on Skolpen Bank in the Barents Sea and Fig. 38b is a double trace, part showing a dense mass of fish on the bottom and part showing a cloud of single fish rising up towards the surface. The fish on the bottom were identified as cod, but, strictly, those in mid-water can only be presumed to be cod. In general, the traces of smaller shoals can be distinguished from those of herring, because they are deeper and the shoal is large, yet still smaller than the cross section of the sound beam. The layers and the mountainous traces are hard to distinguish from their analogs amongst the herring and pilchards, unless it be that the cod layers are deeper. All these records were published by Cushing (1963).

In general, species of fish cannot be distinguished by the shape of their traces on the echo-sounder record. The great diversity of traces is really a function of beam angle, depth, and shoal size as shown in the previous section. Figure 39 is an echo record of whiting in the southern North Sea in autumn at night and should be contrasted with those in Fig. 36; it is a cloud of very small shoals. Superficially, the traces are of the same type, a cloud of spots. That of the herring is in a more strictly defined layer, the spots are smaller and the signal strength is less. So if we add to the character of shoal size, the depth of the trace, the apparent distance apart of each fish or shoal, and the signal strength of the fish or shoal, we have a series of factors which can be used together for descriptive purposes.

Because the shapes of fish traces are governed very much by the directivity pattern of the transducer, and because shoal size and shoal packing are very variable, it is not surprising that it is very difficult to distinguish fish species by their appearance as echo traces. However, given much higher discrimination with narrow beams and short pulse lengths, it might be

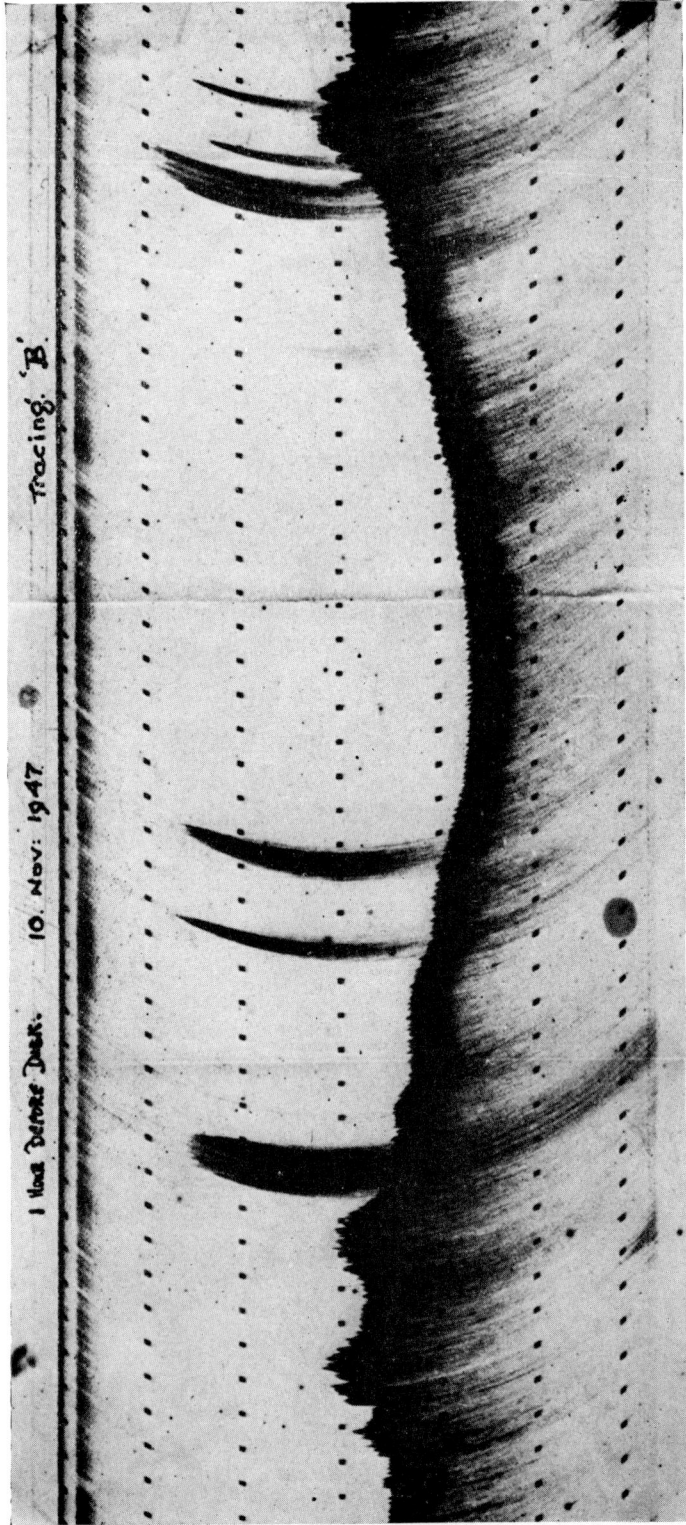

FIG. 37c. Echo trace of a layer of pilchards, south east of the Dodman Point, in Cornwall, in winter (Cushing, 1963).

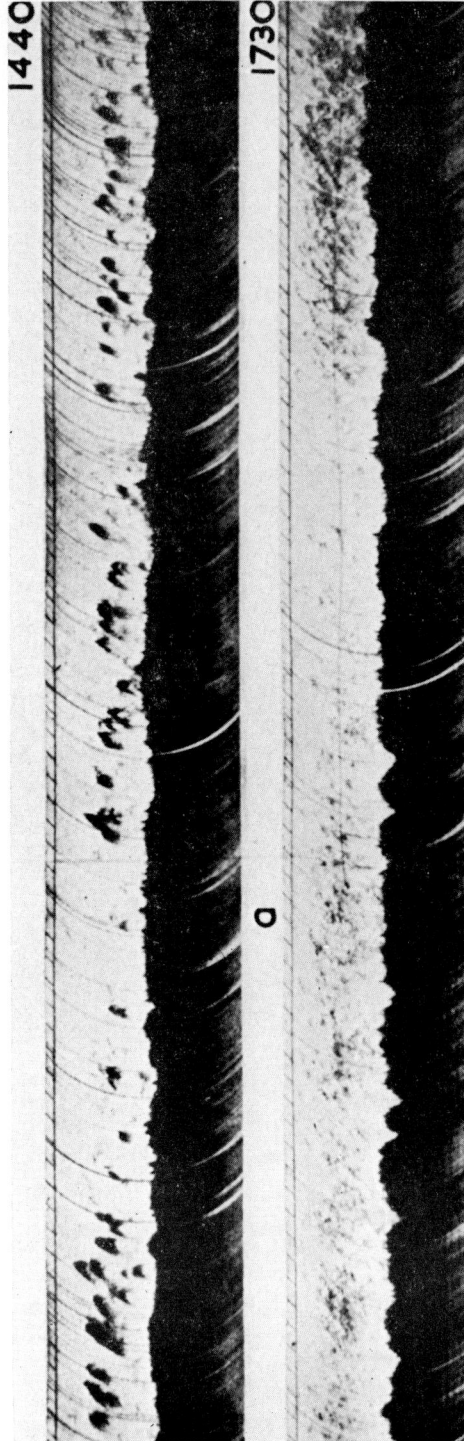

FIG. 38a. An echo trace of small shoals of cod in about 100 fm on Skolpen Bank in the Barents Sea in September (Ellis, 1956).

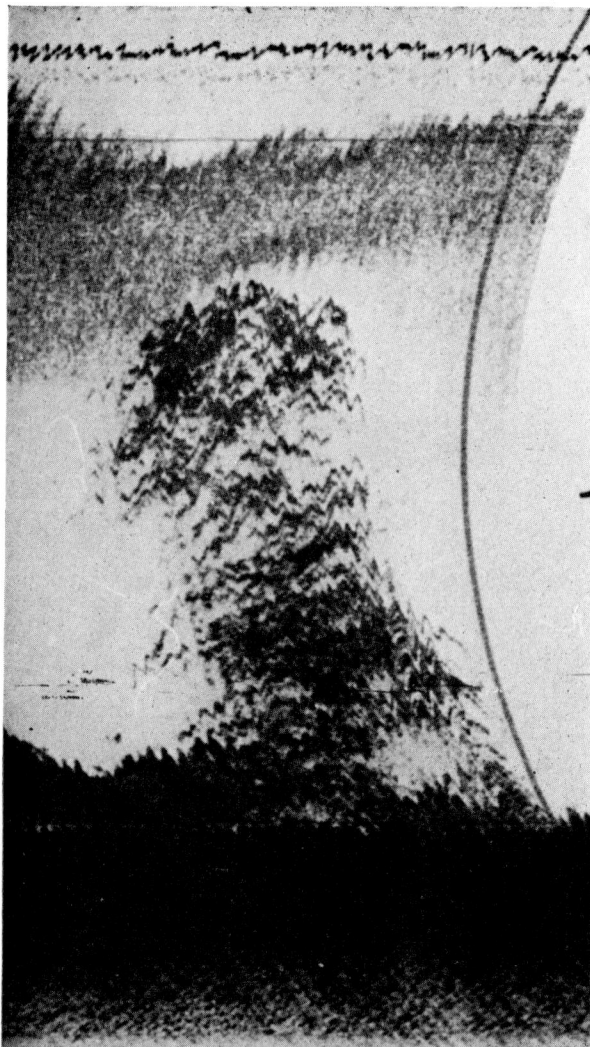

FIG. 38b. A thick echo trace of cod on the bottom, with a cloud of single fish traces rising towards the surface off Bear Island, in the Barents Sea in summer (Cushing, 1963).

possible to use shoal size, shoal packing, and the target strength of individuals to distinguish different fish species. Of these variables, the latter is probably the most valuable because it is dependent upon the sizes of fish and, as will be shown in the next chapter, it is possible that there are differences in target strength between species of the same length.

(d) *Signals from objects other than fish*

Signals from fish are commoner than anything else in the mid-water. But under extreme conditions the echo sounder as commercially marketed can record signals from plankton

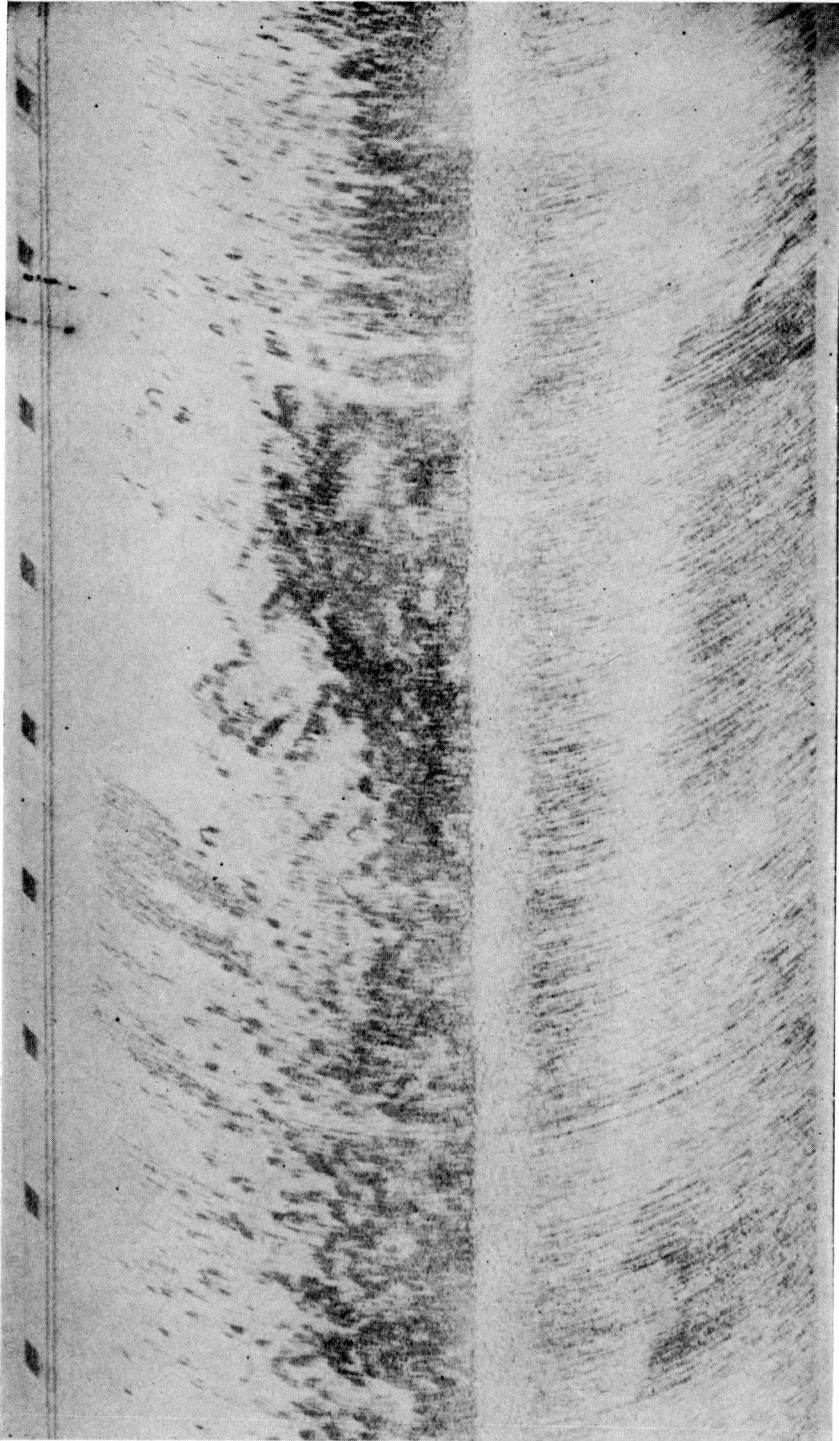

Fig. 39. A difference between echo traces: traces of whiting in winter in the southern North Sea which should be compared with those in Fig. 36.

organisms and from physical structures in the water. In the Baltic are found very sharp density discontinuities and Lenz (1965) has published very pronounced echo records from them; he also showed that signals were received from thermal discontinuities as distinct from density layers caused by salinity differences. The density units across the discontinuities in salinity ranged from 5 to 10 units of σ_t and the thermocline was very sharp, about $0.5°C$ in 1 m. Tveite (1969) has examined echo records from density and temperature discontinuities in the Oslo fjord. Small animals were sampled at 1 m intervals of depth with a pump and large ones with a net towed horizontally for about 1300 m above, below and in the layer; the greatest densities of animals, of either size, were never found within the echo layer. An attempt was made to estimate the differences in acoustic impedance (see Chapter 3) and to correlate these with the theoretical signal; it failed because the observations were taken at 1 m intervals of depth. However, the echo layer must have been due to the physical discontinuities alone, and they are at the same depth night and day, with no diurnal variation as might be expected if animals were causing the layer. Figure 31 shows an echo record from a thermocline in Windermere, in the Lake District, in northern England (Cushing et al., 1955); in this case the temperature difference amounted to about $1.5°$ m^{-1}, but the signals were received from small plankton organisms concentrated at the level of the thermocline. They were sampled with an Ekman water bottle suspended on its side and opened by a frogman. Olsen (1960) found a similar single pulse length layer off the west coast of Newfoundland in July; the layer lay above the thermocline or in the transition layer itself and was probably a layer of animals. Weston (1958) examined a layer at the level of a rather sharper thermal discontinuity in the northern and central North Sea in summer time; he showed that there was not enough acoustic impedance in the discontinuity alone to generate a signal and concluded that the layer was in fact composed of plankton organisms at the depth of the thermocline.

Plankton animals have been detected ultrasonically independently of discontinuities in the water. Cushing and Richardson (1956) recorded a trace of daphnids in Windermere at a density of $27 \, l^{-1}$ in a depth of 16 m; they also recorded a trace of euphausids at a density of $0.66 \, l^{-1}$ at a range of up to 50 m with a transducer firing horizontally in the central North Sea in summer time. Both traces had a noise-like character in that they increased in range as the gain of the receiving amplifier was increased. A more solid target, like a shoal of fish, would be recorded at constant range and, with increased gain, there would be no increase in range. But ordinary echo traces, with no noise-like character, have been recorded more recently by Schröder and Schröder (1964); a layer of zooplankton animals (*Daphnia, Cyclops, Diaptomus, Diaphanosoma*, and rotifers) at a density of $1–10 \, l^{-1}$ was recorded as an echo layer, with an echo sounder more powerful than that used a decade earlier by Cushing and Richardson.

In a later chapter on the Deep Scattering Layer, which is largely concerned with the resonance of air bladders, the layers of fish larvae will be described in more detail. However, Burd and Lee (1951) have shown that a layer of larval pilchards in the western English Channel in early summer was recorded by an echo sounder at a density of 6 larvae m^{-3}. There is a very sharp contrast between the density of fish larvae (6 larvae m^{-3}) and that of plankton animals (600–36,000 larvae m^{-3}) needed to produce an echo trace with the same sort of echo sounder; that used by Burd and Lee was less sensitive than the two used by Cushing and Richardson. Thus, very roughly, euphausids at a density of 600 m^{-3} were the acoustic equivalents of fish larvae at a density of 6 m^{-3}, a difference of two orders of magnitude; on this scale, the difference in size between fish larvae and euphausids is not very great. The real difference is that the little fish larvae have swim bladders which may resonate to the frequency of the transmitted sound.

The air bladders of the post-larval fish are about 1 mm in diameter and may resonate at 30 kHz and the increase in amplitude due to resonance could be considerable (Weston, 1967). The ratio of densities of euphausids to post-larval pilchards needed to yield a signal is of two orders of magnitude which is greater than the possible increase in amplitude due to resonance. Perhaps the larval air bladders were nearer to true air bubbles and did not suffer the restraints in muscle tension and wall thickness expected in the swim bladder of adult fish. So it is indeed possible that the air bladders of the post-larval pilchards were resonating. A similar explanation may account for Northcote's (1964) record of *Chaoborus* larvae in Corbett Lake in Canada, because the larvae have small paired air sacs which may resonate at 200 kHz.

There are certain uncommon situations in which signals can be received from physical discontinuities. A thin layer of silt has been recorded in Lake Huron near the surface analogous to the trace of the thermocline (Cushing, 1963) and Schröder and Schröder (1964) have described veils of turbidity where the Rhine current runs into Lake Constance. Air bubbles from a frogman, from a ship's wake (which is a persistent layer of air bubbles in the sea), and from a ship which has just gone astern, all give intense signals, which were published in Cushing (1963).

Different types of seabed give different echoes, rock a strong one and mud a weak one; in Chapter 3, the differences in reflectivity are given in normal transmission. With a wide angled transducer, such differences in reflectivity are recorded as path lengths received from different angles. Consequently, the extent of the bottom trace beyond the true depth of the seabed varies with the nature of the bottom, long for rock and short for mud. Fishermen use this form of transducer to find the muddy grounds where shrimps live.

But the differences in the thickness of the bottom trace reveal detail in the seabed and some interpretations of distortions of the bottom echo are shown in Fig. 40. In each little picture, the true bottom is shown as a thick hard line and the cone set on the seabed represents the beam angle of the echo sounder. The dotted line shows the distortion of the bottom trace by a particular feature. Figures 40a, b and c show that the echo sounder does not

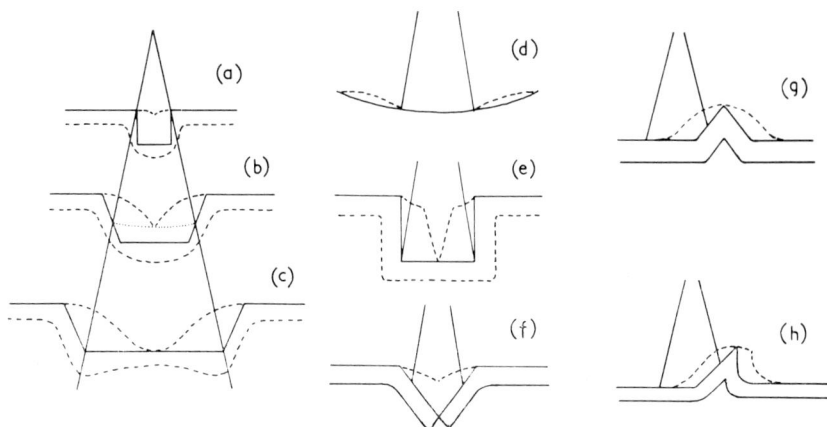

FIG. 40. The distortion which can arise in echo records of the sea bed. The full line shows the true contour and the dotted line represents the distortion; the triangle represents the beam of the echo sounder (Schüler, 1951).

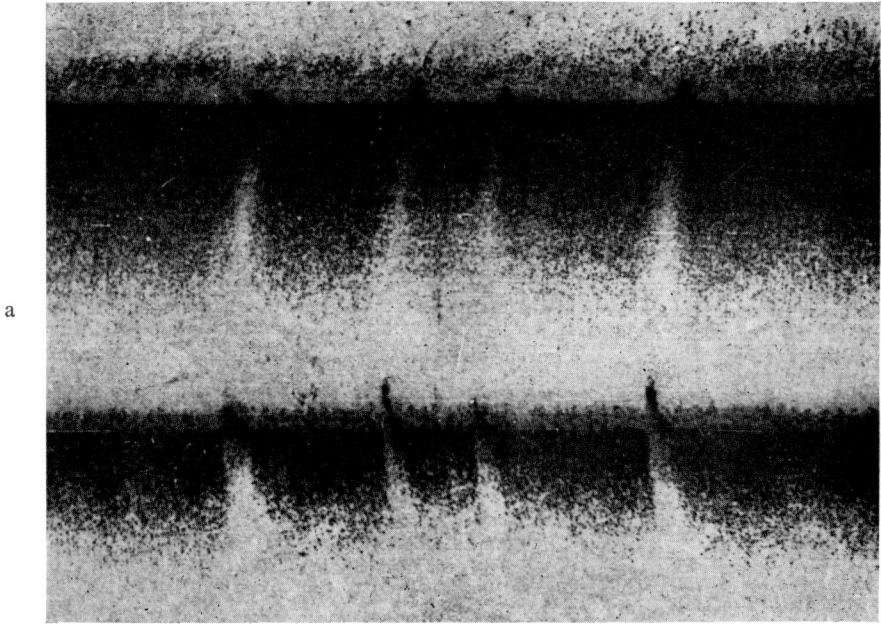

FIG. 41. Distorted traces of the bottom: a, Ridges on the seabed. b, Narrow trenches (Cushing, 1963).

resolve a hole in the seabed less than the effective width of the beam there. Sharp ridges (Figs. 40g and h) are poorly indicated, shallow depressions can become distorted (Fig. 40d), and a trench can appear as a V-shaped valley (Fig. 40e). The dimple in the lower edge of the seabed echo in Fig. 40f betrays the distortion. Bottom traces are shown in Figs. 41a and b; the first shows ridges indicated by gaps in the bottom edge of the trace and the second shows narrow trenches as sharp comet-like extensions of the lower edge of the bottom trace. Such observations are not used by fishermen or by fisheries biologists, but from the trivial interpretations they learn how an echo sounder works.

4. Conclusion

The first need of a fisherman or a fisheries biologist is to detect the presence or absence of fish. Before the recording echo sounder was introduced fish were seen and heard to a limited extent but their presence was often inferred indirectly from the presence of birds or plankton. But their presence is shown directly with the echo sounder and if carefully used the absence of fish can be equally well demonstrated. Most of the mid-water signals are received from fish. It has been shown that plankton organisms at very high densities, fish larvae with air bladders of certain critical sizes, animals or silt laid on thermoclines, very intense physical discontinuities, whales, and dolphins can generate recognizable signals, but this is really a list of special circumstances. The common mid-water signals are from fish, whether in shallow water or in the deep ocean.

The early echo sounders were operated at rather low frequencies, 10–30 kHz and at rather low power; the electrical power input into the transducer was usually less than 100 W. Their modern descendants are much more powerful, using 1–8 kW into the transducer, which is usually constrained into a narrower beam, so the whole system is very much more powerful. The standard echo sounders for work in deep water are operated at 30–50 kHz and for work in shallow water there is a tendency to use frequencies of up to 200 kHz. The advantage of higher frequency is that a much narrower beam can be obtained for the same size of transducer and a much shorter pulse length is possible; so a greater resolution both in angle and in range is obtained. For example, single sprats and single herring can be readily detected at night with the Lowestoft 100 kHz or the Aberdeen 400 kHz machines. Further, at such frequencies the noise received from the sea is very much reduced and so, for the same power output, the system is much more sensitive. However, there is a disadvantage in the use of high frequency and it is that the transmitted power is attenuated to a much greater degree; up to 100 kHz, the loss is not too great; between 100 and 500 kHz, the loss is great enough to preclude the use of such machines except in shallow water; above 500 kHz, it is doubtful whether such frequencies will really ever be used in echo sounders. Thus there is a conflict between the losses due to high absorption and the gain in high resolution and the choice of a particular frequency really depends upon the nature of the compromise.

Fish range in size from 5–150 cm and can be detected as individuals or as shoals depending upon the frequency chosen, which in its turn depends upon the resolution desired and the range needed. In general, bigger fish live in greater depths of water, at least as far as the greatest depths of light penetration. So, if we need to look at cod or hake in depths as great as 360 fm, then a high powered machine (1–8 kW output into the transducer) with a fairly narrow beam (9° × 14°) and a fairly short pulse length of 250 mcsec will resolve most of the animals as individuals, particularly with the high resolution in time of the Kelvin Hughes triggered pen recorder, i.e., the paper recorder is a very fast one. But with a shorter pulse length of 100 msec and a slightly narrower beam and the high time resolution of the Alden recorder, the Lowestoft 100 kHz machine will resolve single sprats and herring at night (and sometimes in daytime). So there is a tendency for low frequency, high powered machines to be used in deep water, with high frequency, moderately powered machines in shallow water, the first being used for big fish and the second for little fish.

The detection of fish has been considered so far as recording their presence or absence; in other words the recorder has been used as indicating the presence or absence of fish without any quantitative evaluation of the signals. The next step is to determine how far the quantitative use of the echo sounder can be pressed, but before it is described in detail, a brief statement of some of the formal acoustic problems is given in the next chapter.

CHAPTER 3

SOME ELEMENTARY ACOUSTICS

1. *Introduction*

So FAR the use of the echo sounder has been described with the least use of the physical principles involved. Fishermen, fisheries biologists, and navigators employ the machine as a continuous indicator of the depth of the sea or of the presence or absence of fish in the mid-water. In the last chapter many of the subtleties of the echo sounder record were described but they were interpreted in terms of simple geometry. It is possible to use echo sounders in a quantitative manner, and this forms the basis of the relative and absolute methods of estimating abundance to be described later.

Before the quantitative use of echo sounders can be fully described, some account of the elementary acoustic principles is needed. Some of the most valuable work in underwater acoustics is represented in the sonar equations which express the changes in quantities at a receiving transducer for a number of conditions. The purpose of this chapter is to introduce the terms needed to understand the sonar equations and to describe the measurements made of the signal received from fish under standard conditions. For a full development of acoustic theory in underwater sound, other sources should be consulted, for example, Urick (1967).

2. *Sound waves in the sea*

A sound wave, being a compressional one, is longitudinal, which means that particles are displaced back and forth in an elastic medium in the direction of propagation. In the sea, sound travels at about 1500 m sec^{-1}, but this quantity varies slightly according to the temperature, salinity, and pressure (see p. 76). Although the particles of water are not translated, there is transmission of energy by this movement back and forth in compressional waves and the power transmission per unit area is the intensity. Intensity decreases from the source as the inverse square, because the wave is a spherical one and the energy transmitted per unit area decreases as the wave spreads with range; it also decreases due to absorption by the medium and the sum of spreading and attenuation is called the transmission loss.

Like light waves, sound waves are reflected from physical discontinuities; sound travels through different materials at different speeds and they are reflected or refracted as they reach them. They are also diffracted and interferences between waves can become important in the measurement of the target strength of fish and in the construction of transducers. A transducer is measured in wavelengths at the frequency to which it is tuned, e.g., 3λ by 2λ, 6λ by 4λ, if it is rectangular, or 2λ in diameter if circular. It will be shown below that the beam angle is a function of the number of wavelengths across the face of the transducer; the bigger the transducer, the narrower the beam at a given frequency. It follows (because $\lambda = cf^{-1}$ where c is the speed of sound in cm sec^{-1} and f is frequency in Hz) that with higher frequencies, transducers are smaller for the same beam angle, or that for the same size of transducer, a narrower beam angle may be obtained at higher frequencies. In the range 10–60 kHz, transducers may be made of nickel and employ the magnetostrictive effect, but particularly at higher frequencies they are usually made of ceramic and employ the piezoelectric effect: a pulse of electrical energy is converted to one of acoustic energy by a physical translation of the transducer material by either of these effects. Transducers are generally constructed to be mechanically resonant at their working frequency. The quality factor (Q) describes the damping of a transducer resonance when it is energized: a high Q means that the damping is low and a low Q that the damping is high. The higher the Q, the more cycles of the resonant frequency are required to reach peak energy. Thus, for the same Q a higher frequency transducer can respond to shorter electrical pulses more effectively and for the same size of transducer an increase in frequency means narrower beam angles and shorter pulses or greater angular resolution and greater range resolution. The disadvantage is that the absorption coefficient, a, increases with frequency and so the maximum range is less, as indicated at the end of the last chapter. But in the sea, water noise decreases sharply with frequency (to about 100 kHz, where it becomes thermal noise) and so some of the losses due to absorption are countered by gains in increased signal-to-noise ratios. Obviously, there are a number of factors to be considered and any equipment constitutes a compromise between conflicting needs. It will be shown below that the compromise can be evaluated quantitatively by means of the sonar equations. In order to proceed to them, we now return to the sound waves and show how the pressure changes are generated by them and how the energy flow can be combined in terms of transmission, spread, directivity, and signal received.

A little more formally, the particle motion in a fluid is parallel to the direction of propagation and because the fluid is elastic, the back and forth motion of the particles causes pressure changes. The pressure generated in a plane wave depends upon the velocity of the particles:

$$p = \rho c v \tag{2}$$

where p is pressure in dyn cm^{-2}; ρ is the density of the fluid in g cm^{-3}; v is the velocity of the particles in cm sec^{-1}; c is the velocity of sound in the sea in cm sec^{-1}.

In the direction of propagation, energy will flow across unit area and the flux of energy per unit time across it is the intensity. Then

$$I = p^2/\rho c \tag{3}$$

where I is the instantaneous intensity in erg cm^{-2} sec^{-1}.

During a time period, $I = \overline{p^2}/\rho c$ where $\overline{p^2}$ is the average pressure during that period. The intensity in W cm^{-2} is given by

$$I = (\overline{p^2}/\rho c)\ 10^{-7} \tag{4}$$

However, the unit of intensity used in underwater sound is that of a plane wave with an r.m.s. pressure equal to 1 dyn cm^{-2} (or 1 mcbar) and this is a standard pressure to which all measurements are referred (it is sometimes expressed as "*ref* 1 mcbar" or "*ref* 1 dyn cm^{-2}"). Because there are considerable changes in magnitude of the variables, the decibel (dB) is used; it is a convenient unit, being a logarithmic ratio of intensities (or of pressures).

$$N = 10 \log I_1/I_2 \ \text{dB}$$

where I_1 and I_2 are two intensities and N is expressed in decibels. It is, at first, a little difficult to remember what a ratio in decibels really amounts to, however easy it is to use in the sonar equations. The following brief table may serve as an aide-memoire:

dB	Intensity ratio	Pressure ratio
1	1.26	1.12
3	2.00	1.41
5	3.16	1.78
6	4.00	2.00
10	10.00	3.16
20	100.00	10.00
40	10,000.00	100.00

The source level S, is a measure of the power output on the acoustic axis of the sound beam at unit distance from the transducer;

$$S = 10 \log F \tag{5}$$

where F is the sound intensity on the acoustic axis at unit distance from the transducer.

When the sound is transmitted from the transducer the spherical wave is constrained into a conical beam. The transducer may be considered as a number of point sources from each of which emanate the spherical waves. Normal to the transducer face all the waves are in phase and so this is the direction of maximum intensity, but at any angle to the axis the spherical waves interfere and, as a consequence, intensity decreases with angle from the axis. From the diffraction equation, the directivity pattern function, b, in one plane for a rectangular transducer is given by:

$$b = \left[\frac{\sin[(\pi a'/\lambda) \sin \theta]}{(\pi a'/\lambda) \sin \theta} \right]^2 \tag{6}$$

where a' is the length of the transducer in cm; θ is the angle from the acoustic axis; and λ is the wavelength of the sound in cm.

The directivity pattern function is that for a length of transducer a'; for the other dimension a'', the pattern may differ and so the beam is not symmetrical; if $a' = 3\lambda$ and $a'' = 2\lambda$, the beam in the dimension a' is the narrower. Consequently the larger the transducer, at a given frequency, the narrower is the beam.

In a circular transducer, the beam is symmetrical, and the directivity pattern function is given by:

$$b = \left[\frac{2J_1[(\pi a'/\lambda) \sin \theta]}{(\pi a'/\lambda) \sin \theta} \right]^2 \tag{7}$$

where a' is the diameter of the transducer in cm, and J_1 is the first order Bessel function.

The sound intensity at any point in the beam is given by:

$$I_{(r,\theta,\phi)} = F.b_{(\theta,\phi)}/r^2 \tag{8}$$

where ϕ is the angle in azimuth. Thus the effects of angular distribution and spreading in range, r, due to the inverse square law are described. The sound level, L', can then be expressed:

$$L'_{(r,\theta,\phi)} = S + 10 \log b_{(\theta,\phi)} - 20 \log r \tag{9}$$

where S is the source level in dB ref 1 mcbar. However, there is also a loss due to absorption, largely because ions of magnesium sulfate dissociate under the pressure of the sound wave. It has been shown that absorption increases with the square of the frequency, which has the important consequence that frequencies above 500 kHz cannot generally be employed successfully at sea because the useful range becomes so much diminished. Absorption is also dependent on temperature and upon pressure. Nomograms have been constructed to calculate the total transmission loss due to spreading and attenuation (or absorption with range) for a variety of conditions. Effectively the transmission loss A,

$$A = a.r/1000 \tag{10}$$

where a is the attenuation coefficient (dB km^{-1}), and r is the range (m).

Then the one way transmission loss, H, is given by:

$$H = A + 20 \log r \tag{11}$$

in which the effects of spreading and attenuation are added.

Then $$L' = S + 10 \log b - H \tag{12}$$

The velocity of sound in the sea, c, varies:

$$c = (E/\rho)^{\frac{1}{2}} \tag{13}$$

where E is the bulk modulus, or the modulus of elasticity. Both density and elasticity are functions of temperature, salinity, and depth and so velocity can be expressed:

$$c = 1400 + 4.9t - 0.044t^2 + (1.32 - 0.01t)S + 0.018 (1 - 0.01t)z \tag{14}$$

where t is temperature ($^{\circ}$C); S is salinity (ppm); z is depth (m).

Because temperature and salinity vary with depth, so does sound velocity. Temperature usually decreases with depth, salinity tends to increase with depth and so there is a mid-ocean channel at moderate depth at which sound velocity is least. Indeed, this channel can be used for trans-oceanic signalling because the vertical distribution of velocity creates a wave guide. Such effects do not affect echo sounding very much because the beam crosses the layers normally, but sonar operations are often markedly hampered by them. However, the detailed effects of refraction will not be described in this chapter.

The material of a target governs its reflectivity. The coefficient of reflectivity, R, is given by:

$$R = \left[\frac{\rho_2 c_2 - \rho_1 c_1}{\rho_2 c_2 + \rho_1 c_1}\right]^2 \tag{15}$$

where $\rho_1 c_1$ is the acoustic impedance of the medium and $\rho_2 c_2$ is that of the target. So, reflectivity depends on the difference in impedance across the discontinuity between medium and target. An air–water boundary has sharp differences in acoustic impedance and so the reflection coefficient is high (99.88%).

The reflectivity of the seabed is acoustically complex, but Hashimoto (1953) quotes the following vertical bottom reflection coefficients: mud 28%; sand, 42%; rock, 63%. Urick (1967) shows how the scattering strength, in dB, varies with frequency and with grazing angle. The scattering coefficient increases with frequency unless the seabed is heavily dissected, and it increases with grazing angle. The most important point is that the bottom roughnesses are probably more important than its material; the problem is confused because muddy bottoms tend to be smooth and rocky bottoms rough; a fuller analysis may be found in Urick's book. Breslau (1967) measured the loss of signal in the seabed (at 12 kHz) and has correlated it with various characteristics of the sediments, i.e., positively with porosity and the percentage of fine silt or clay and inversely with median grain size and the sorting coefficient. With his methods it should become possible to chart areas of sand, gravel and silt directly with an echo sounder.

3. *The sonar equations*

Any measurement with an underwater acoustic system is made against a background of noise. There are three sources of noise, instrumental noise, environmental noise (like flow noise, ambient water noise), and reverberation (as the transmitted wave returns by many and devious paths to the receiver). Such sources of noise are frequency dependent; for example, the ambient water noise declines with frequency to below thermal noise at about 100 kHz. There is more noise at higher wind strengths and there are sources of noise from the sea surface and from various sorts of animals like croakers and snapping shrimp. At 10–20 kHz Hashimoto and Maniwa (1956) have distinguished between the noises made by spiny lobsters and blue crabs walking on the sand and those made by sunfish and filefish swimming. Biological noises are not randomly distributed because their sources are in localized patches. Whatever the source of noise it should be measured because, as will be shown below, signals from the desired targets can be noise-limited; with echo sounders, the prime source of noise comes from the ship's propellers and it decreases with depth of water so the noise level has to be monitored fairly frequently. In general terms, noise decreases with increased depth of water, increased frequency (to 300 kHz), and with a decrease in engine noise.

There are three forms of parameter in the sonar equations, determined by the equipment, the medium, and by the target. Obviously the source level, S, and the directivity characteristics are properties of the equipment, but so is the level of instrument noise (or self-noise). The medium governs the transmission loss, H, and the levels of reverberation and ambient noise. The target strength expresses the signal reflected or scattered back from a target at unit distance from the target in the direction of the transducer. One sonar equation defines the detection threshold:

$$S - 2H + T - (NL - D) = DT \qquad (16)$$

where S is the source level in dB *ref* 1 mcbar (1 dyn cm^{-2}); T is the target strength in dB; and DT is the detection threshold in dB. Thus the detection threshold, as a signal-to-noise ratio, is defined in terms of the target to be detected, the equipment, and the characteristics of the medium. Then

$$S - 2H + T = NL - D + DT \qquad (17)$$

This formulation is convenient because the echo level occurs on the left hand side of the equation and the noise components which may include reverberation on the right. A shorter way of writing this equation is:

$$EL = S - 2H + T \qquad (18)$$

where EL is the echo level above a conventional signal-to-noise ratio. This is the form of sonar equation, which is employed in the quantitative use of the echo sounder. Thus the signal observed at any one time is considered to be a function of three parameters; the source level, S, is a characteristic of the equipment; the one way propagation loss, H, is governed by the medium and by the frequency used; and the target strength, T, describes the acoustic magnitude of the target in the sea.

A useful way of comparing different echo sounders is given by the performance figure:

$$S - (NL - D)$$

which is effectively the difference in decibels between the source level and the noise level (or reverberation level), taking different beam angles into account. This figure of merit is limited to a comparison of effective power, perhaps including differences due to frequency. But not all the useful characteristics of the equipment are taken into account; for example, the resolving power in angle or in range is excluded.

Many applications of the sonar equation are modified by pulse length, particularly if the target, like a submarine, is large as compared with the transmitted pulse. Individual fish are usually much smaller than the transmitted pulse and so the sonar equations can be used directly without modification on distributions of individual fish. With large targets like fish shoals, the assumption that the received pulse is of the same length as the transmitted one may no longer hold. However, as will be shown below, the best solution to the problem of fish shoals may be to reduce them to a mass of discrete individuals, signals from each of which can be summed.

4. The target strength of fish

(a) The target strength of a sphere

The power intercepted from an incident sound wave by a rigid sphere of radius a is $\pi a^2 I_i$, where I_i is the incident intensity. At r m from the sphere, the intensity of the reflected wave I_r, will be the ratio of the incident intensity to the area of a sphere of radius r, or

$$I_r = \frac{\pi a^2 I_i}{4\pi r^2} = I_i \frac{a^2}{4r^2} \qquad (19)$$

At the reference distance of 1 m:

$$\left.\frac{I_r}{I_i}\right|_{r=1} = \frac{a^2}{4}$$

Then the target strength of the sphere is given by:

$$T \equiv 10 \log \left.\frac{I_r}{I_i}\right|_{r=1} = 10 \log \frac{a^2}{4} \qquad (20)$$

So the ideal sphere of radius of 2 m has a target strength of 0 dB. That of a submarine has a positive value, whereas that of an individual fish is negative. Thus, target strength is the fraction of incident intensity returned by a target at a distance of 1 m from its "acoustic center".

A definition analogous to that of target strength is that of scattering cross section, σ:

$$\sigma = 4\pi \left.\frac{I_r}{I_i}\right|_{r=1} \quad ; \quad T = 10 \log \left.\frac{I_r}{I_i}\right|_{r=1}$$

$$\therefore \ T = 10 \log \sigma/4\pi \qquad (21)$$

So, the sphere of a radius of 2 m has a target strength of 0 dB and a scattering cross section of 4π, which is a unit of area.

Target strengths are calculable for certain rigid shapes of different sizes. For rigid spheres much smaller than one wavelength, the scattering cross section, or target strength, varies with the fourth power of the frequency but when the target is about the size of one wavelength, in the so-called middle zone, marked oscillation in target strength occurs. The scattering from a fluid sphere is calculable (Anderson, 1950), but that from a fish body is more complex. The shape of a fish approximates to an ellipsoid, but the body is composed of swim bladder, backbone, and scales and they each respond differently.

(b) *The impedance of fish flesh*

Fish flesh differs only slightly in density from sea water. The following table gives the reflection coefficient of air and fish flesh in sea water:

Material	Reflection coefficient (%)
Air	99.88
Fish flesh	1.2–6.0

Shibata (1970) has reported a number of measurements of the speed of sound in fish using a form of flaw detector at 2 MHz; the measurements may be expressed as sound velocity (msec^{-1}), acoustic impedance (ρc), absorption (dB, cm^{-1}), and as reflection coefficient (%). The following table summarizes his results:

TABLE 5. REFLECTION COEFFICIENTS OF FISH FLESH IN VARIOUS SPECIES

Fish	Coefficient (%)
Clupea pallasi (herring)	2.2
Scomber japonicus Houttuyn (mackerel)	3.3
Seriola dorsalis Gill (yellowtail)	2.5
Anoplopoma fimbria (gadoid)	1.2
Arctoscopus japonicus Steindachner (sailfin sandfish)	2.3
Carassius auratus Linnaeus (goldfish)	3.8
Onchorhynchus gorbuscha Walbaum (pink salmon)	4.6
Thunnus obesus Lowe (big-eye tuna)	6.0

Shibata also showed that the loss in scaled flesh was 1.63 that of unscaled flesh measured in dB cm^{-1}. Mahrous and Cushing (1956), with some very crude measurements, suggested that the speed of sound in fish bone was twice that in scales. Hence the fish as a target is a complex one, with at least three components of different sizes and impedance. Shibata (1970) suggests that the reflection coefficient of pelagic fish (3.5%) tends to be greater than that for demersal fish. On the basis of sound velocity measurements on goldfish, sticklebacks, guppies, mackerel, and sardines, he has partitioned target strength measurements as follows:

$$T = 10 \log (\sigma_f + \sigma_{ab} + \sigma_v)/4.10^2 \tag{22}$$

where $\sigma_v = 0.015\pi \, (0.13)^2 \, L^3/\lambda$

$\quad\quad \sigma_{ab} = 0.066\pi \, (0.25)^2 \, L^3/\lambda$

$\quad\quad \sigma_f = 0.151\pi \, (0.092)^2 \, L^3/\lambda$

where σ_v is the scattering cross section of the vertebrae; σ_{ab} is the scattering cross section of the air bladder; σ_f is the scattering cross section of the fish flesh; and L is length (cm).

An interesting point is that the polar diagram of the living fish appears to be less variable than that for dead fish. Another important point is that the target strength of the vertebrae is 12.5 dB down on the air bladder; further, the differences due to the three components tend to blur differences in frequency.

Thus, as compared with an air–water interface, the flesh–water interface is of low reflectivity, but that of bones and scales is much higher. Hence, fish without swim bladders, like mackerel, dogfish, or some tuna can still be detected quite easily with echo sounders.

(c) Measurements of target strength

The first measurements were made by Smith (1954) who determined the target strength of a sea bass at rather short range. A detuned transducer was used at frequencies from 8–30 kHz and the target was moved in the beam to find maximum signal, so even at the short ranges used (about 1 m) proper estimates of target strength were obtained. There was no trend with frequency for a sea bass in side view, but a drop of 5 dB with increased frequency in end view. The target strengths of scup (*Stenotomus chrysops* Linnaeus), squid, and shrimp declined with frequency; Smith thought that the targets were in the middle zone of frequency, where such variation might be expected. It was shown clearly that the target strength of shrimp was not enough to account for the signals from the Deep Scattering Layer unless the numbers per unit volume were extraordinarily high. Hashimoto and Maniwa (1956) made a number of measurements of "reflection loss" (which is effectively target strength) from a number of dead fish, including some at high frequency (200 kHz). Hashimoto and Maniwa (1956) developed a formula for "reflection loss"; when $L' \leqslant \lambda$ the reflection loss, RL', is given by

$$RL' = 20 \log \left(\frac{\lambda}{2r_c}\right)^{\frac{1}{4}} \cdot \frac{100}{L} \cdot \frac{2}{\Delta\rho} \cdot \frac{1}{K} \tag{23}$$

where r_c is the radius of curvature (cm); $\Delta\rho$ is the difference in density between fish body and sea water; K is a shape constant ($\simeq 1.0$ for the dorsal aspect of mackerel, frigate mackerel (*Auxis thazard* Lacépède), and horse mackerel; Hashimoto and Maniwa, 1956). This formula was developed by Hashimoto (1953) on the basis of the return of sound from a cylindrical solid. For two fish species, the reflection loss in dB, is reduced with frequency:

Species	Length (cm)	Frequency (kHz)				
		28	100	200	300	400
Sardine	19.4	52	46	44	41	39
Horse Mackerel	15	50	42	41	37	33

Hashimoto (1953) made a number of observations on dead fish and suggested that reflection loss decreased with volume of the fish and the square of the frequency. The first rule was confirmed about a decade later with the measurements made by Midttun and Hoff (1962) and Cushing *et al.* (1963); but the second rule was not confirmed by Haslett when the work was repeated in 1969. Hashimoto and Maniwa (1958) examined the effect of the swim bladder and suggested that there was no difference in signal if the swim bladder was filled with water. From experiments on dead fish with artificial swim bladders, Cushing and Richardson (1955) suggested that half the signal from fish was received from the swim bladder. With more precise measurements made on a rigidly suspended fish in a tank, Harden Jones and Pearce (1958) confirmed this suggestion, so denying the conclusion of Hashimoto and Maniwa (1958); they also showed the distribution of signal from all aspects of a fish and that from the side of the fish was shown to be ten times greater than that from the back (Fig. 42). Shibata (1970) also made careful measurements of the contributions of the air bladder; he found that it comprised 52% in dorsal aspect at 50 kHz and 48% of 200 kHz. In other aspects, ventral, head or tail, the proportion contributed by the air bladder was less. Harden Jones and Richardson *et al.* (1959) showed that dead fish suspended on a nylon line could be measured in target strength with reference to an 8-in. trawl

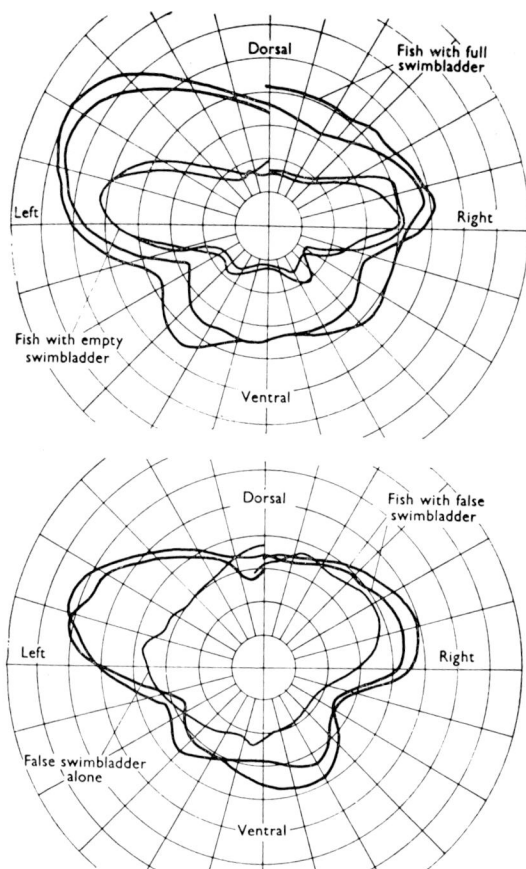

FIG. 42. The distribution of signal amplitude in angle about a fish body, in dorsal, lateral and ventral aspect (Harden Jones and Pearce, 1958).

float, suspended two pulse lengths from the fish on the same line. He constructed a diagram of signal on range from transducer to noise level, which is effectively a pictorial formulation of the sonar equation (Fig. 43). The figure illustrates at one and the same time the decline of received signal with depth or range on the axis of the beam and the increase of received signal with size of fish. The noise level represents the maximum range of the system. This figure was the first to show how quantitative echo sounding for fish might develop. There were two conclusions from this work; first, that a 70 cm cod yielded the same signal as the trawl float and, secondly, that bigger signals were received from bigger fish.

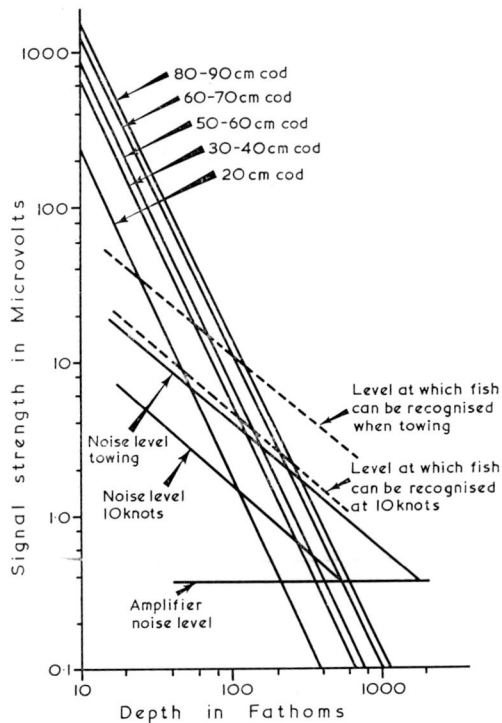

FIG. 43. Signals from fish of different sizes calculated at a series of depths to the noise level (Harden Jones, in Richardson *et al.*, 1959).

The same result emerged from a series of measurements made on dead fish with artificial onazote air bladders at sea, in dorsal aspect, in a dock and in a river (Cushing *et al.*, 1963); although the variance of the measurements was high, it was shown that the target strength of the fish, mainly cod, increased roughly as the volume of the fish. Another series of measurements made by Midttun and Hoff (1962) showed the same result with much reduced variance; they worked at rather longer ranges with freshly killed fish which had been kept at the depth of observation for some days when alive, so the swim bladder was intact and of the right volume. An interesting point from this work was that some evidence showed that the swim bladder generated its own directivity pattern. Later work by Midttun (1966) has suggested that coalfish might be discriminated from cod by the different durations of signal for the same signal strength and this has been shown by Midttun and Nakken (1971). Mr. Bill Woods (personal communication) has told me that he has been able to discriminate individual coalfish from cod by their shorter signals on the paper record. Shibata (1970) has shown that the directivity of the air bladder of the goldfish in dorsal aspect deviated 10° from the vertical. Figure 44 shows the results of Midttun and Hoff (1962) combined with the best of those of Cushing *et al.* (1963) and those of Harden Jones and Pearce (1958). Target strength is shown to increase nearly as the volume of the fish. Recently Clayden (1969) has shown that the relationship can be extended to small fish of 2–15 cm at a frequency of 100 kHz.

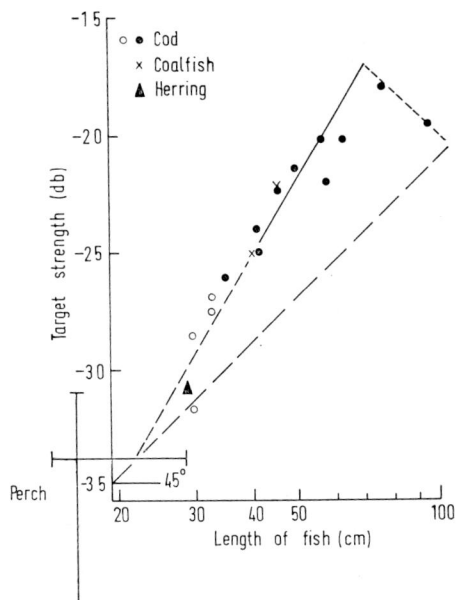

FIG. 44. Dependence of target strength (dB) upon length of fish (in logarithms) (Cushing, 1964).

Haslett (1962 a, b, c, d) made many experiments on model fish and small fish in tanks at high frequency. He also concluded that target strength increased with size of fish, but that in the middle zone of frequency (where $4 \leqslant L/\lambda \leqslant 100$) there was much increased variability, due to interferences between waves at the surface of a target about the size of one wavelength. Haslett suggested that his data from models, made from rubber and from perspex, were fitted by a complex oscillatory curve through the middle zone of frequency. Although this is theoretically correct, Love (1971a) has suggested recently that the effect of interference in the middle zone was much less than had been expected. Love's measurements were made on live fish which were anesthetized and mounted by monofilament line on a rotatable frame one yard from the transducer (which is perhaps a little close) and signals were measured initially from the side of the fish. Figure 45 shows the relationship between σ/λ^2 and L/λ for six species and the relationship is a clear one. The maximum side aspect signals were available for yellowfin tuna, yellowtails (*Seriola dorsalis*), crappies (*Pomoxis annularis* Rafinesque), large mouthed bass, mackerel, beryx, plaice, flounder, *sticklebacks, guppies, minnows, banded killifish (*Fundulus diaphanus* Le Sueur), black and white crappies,

*The plaice and flounder were measured in a morphologically dorsal aspect.

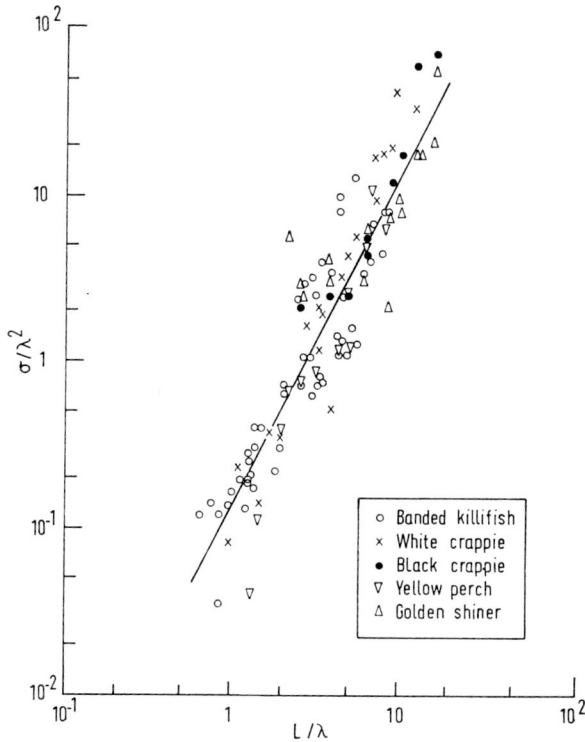

FIG. 45. Dependence of scattering cross section (σ) of fish in lateral aspect (normalized to σ/λ^2) on L/λ, where L is the length of the fish (Love, 1971, a, b), for five species of fish.

yellow perch (*Perca flavescens* Mitchell), and eastern golden shiner (*Notemigonus cryso-leucas* Mitchell). Figure 46 shows the relationship for side aspect:

$$\sigma/\lambda^2 = 0.064 \, (L/\lambda)^{2.28} \tag{24}$$

and

$$T_s = 22.8 \log L - 2.8 \log \lambda - 22.9 \text{ (from } 1 \leqslant L/\lambda \leqslant 130)$$

where L and λ are measured in cm.

Differences between experiments, between fish species, and between bladdered and bladder-less fish are much less than the effect expressed in eqn. (24).

He has also published the results of his measurements of target strength in dorsal aspect. Observations were made at 12, 15, 25, 40, 60, 100, 150 and 200 kHz on anchovies (*Anchoa mitchilli* Valenciennes), menhaden (*Brevoortia tyrannus* Latrobe), goldfish (*Carassius auratus*), striped killifish (*Fundulus majalis* Walbaum), mummichogs (*Fundulus heteroclitus* Linnaeus), Atlantic silversides (*Menidia menidice* Linnaeus), black trout (*Pomoxis nigromaculatus* Le Sueur), and spotted sea trout (*Cynoscion nebulosus* Cuvier). For 16 families, in a range of frequency from 8–1480 kHz, the following relationship was established:

$$\sigma/\lambda^2 = 0.041(L/\lambda)^{1.94} \tag{25}$$

and

$$T_D = 19.4 \log L + 0.6 \log \lambda - 24.9 \tag{26}$$

where T_D is the target strength in dorsal aspect; and L and λ are measured in cm.

The relationship is valid in the range $0.7 \leqslant L/\lambda \leqslant 90$; such a relation presents scattering cross section as a function of length for a constant wavelength. Similarly dependence of σ/L^2 on L/λ expresses scattering cross section as a function of the reciprocal of wavelength for a constant length. If the data are averaged by one third octaves (i.e. by the square root of the frequency difference) in the dependence of σ/L^2 on L/λ, there is a minimum in side aspect at $L/\lambda = 7$ and in dorsal aspect at $L/\lambda = 14$. Very roughly the radius of curvature in side aspect is twice that in dorsal aspect; the formulation put forward by Hashimoto and Maniwa suggests that scattering cross section depends directly on frequency and inversely on the radius of curvature.

McCartney and Stubbs (1971) measured the target strengths of some living fish within a plastic container and within a ring hydrophone. In dorsal aspect, the following relationship was established:

$$T = 24.5 \log L - 4.5 \log \lambda - 26.4 \tag{27}$$

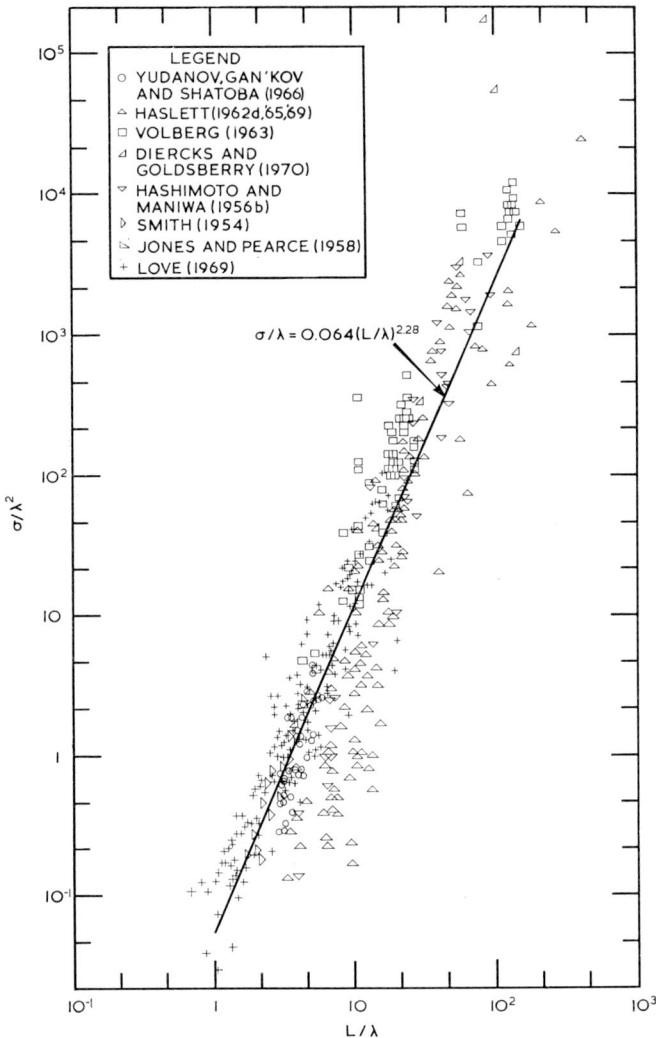

FIG. 46a. Dependence of σ/λ^2 on L/λ for many data on sixteen species (lateral aspect) (Love, 1971, a, b).

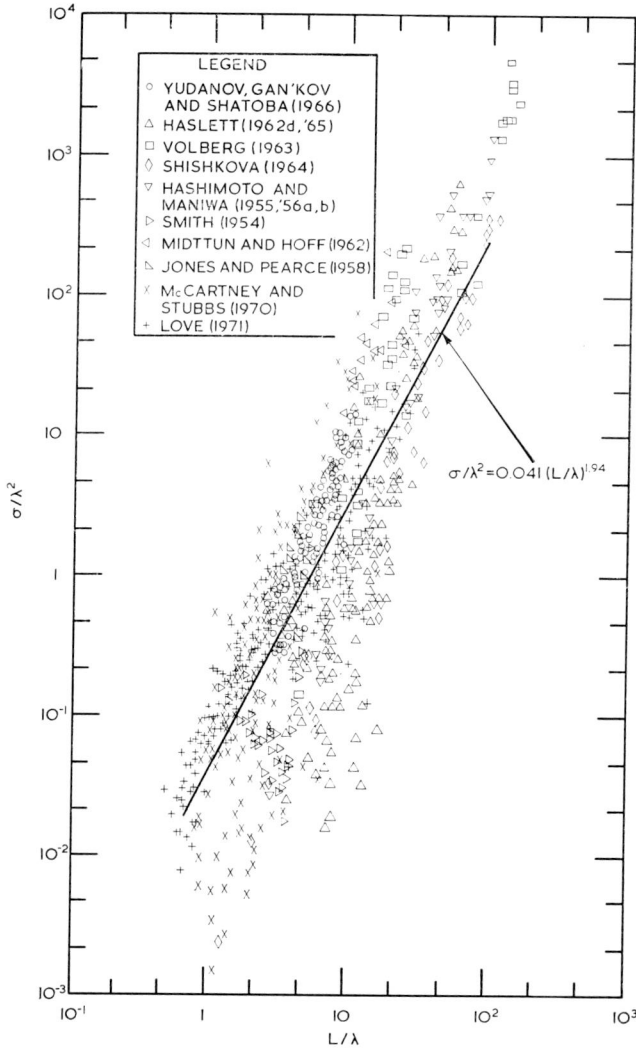

FIG. 46b. Dependence of σ/λ^2 on L/λ for many data on sixteen species (dorsal aspect) (Love, 1971, a, b).

In the form σ/λ^2 on L/λ which expresses the dependence of scattering cross section on fish length, the standard error of the slope was 1.4 dB/decade. The authors write: "there are more points above and close to the line than below it, suggestive of multiple scattering with broad maxima and deep interference minima". So, differences due to radius of curvature are less important than those due to the size of fish.

The most important result from this work is that although there is an increase in variance about the regression in the middle zone of frequency, there is no evidence that the data could be better fitted by an oscillatory curve of the type postulated by Haslett. However, Haslett (1962) has also shown that the signal originates from various parts of the fish's body, backbone and head as well as from the swim bladder and the external surface. So the effective middle zone for a given frequency may well be spread over a wide range, as L (in L/λ) must take different values for different parts of the body. There are two consequences of these valuable observations, first, that the relationship between target strength and size of fish can be extended to most sizes of fish and, secondly, that there may be differing relationships for different species of fish, in side aspect, at least.

As noted above, Hashimoto and Maniwa (1956) suggested that the target strength increased with frequency, as perhaps it should do so, theoretically. Recently, however, Haslett (1969) has repeated this work and found only a slight increase of dB/decade in frequency. Shibata (1970) has summarized his experiments at 28, 50 and 200 kHz in the following equations:

$$T_D = 28 \log L - 8 \log \lambda - 67.4 \text{ (dorsal aspect)} \tag{28a}$$

$$T_s = 25.7 \log L - 5.7 \log \lambda - 63.8 \text{ (lateral aspect)} \tag{29b}$$

where L and λ are expressed in cm.

Hence within the range of frequencies that might be used (30–500 kHz), there is an increase in target strength with frequency, but it is small. If the middle zone in frequency is effectively spread because the reflecting parts within a fish have different "lengths" at one frequency, perhaps the expected trend of target strength with frequency is minimized for the same reason.

Fish cannot be identified as species, acoustically. However, signals from cod and coalfish may be distinguishable by the differing directivities of their swim bladders as noted above. One would expect that the average number of transmissions on a target crossing the beam at a given range would be a function of target strength; differences in swim bladder directivity would alter the effective beam of the equipment and so the ratio of target strength indicated from the number of transmissions to that from the signal amplitude would depart from unity as shown by Midttun and Nakken (1971). Love's work suggests that there may be differences in target strength for different sizes of fish (in side aspect) between species. This is to be expected because the magnitude of signal reflected from a body depends to some extent upon its radius of curvature and fish bodies of different species differ markedly in this particular way. So, although fish species cannot be identified acoustically at the present time, there is a possibility of making some sort of discrimination in the future. Cushing and Richardson (1955) using an echo sounder which operated at three frequencies showed differences in received signal from cod and herring irrespective of size; at the time, the results were interpreted as being the consequence of different middle zones for the two fish species but perhaps a more general conclusion would be that fish of different species may really have different target strength characteristics. Cushing and Richardson also showed that there were differences in variance in received signal between cod and herring; that of cod was greater than that of herring, as might be expected, because the cod are more dispersed within the sound beam.

From the sonar equations, set out in section 3, measurements of target strength are needed before any quantitative use can be made of the echo sounder. Enough measurements have been made to show that target strength increases as the area or as the volume of the fish and, because the fish body is soft and acoustically complex, it is possible that this relationship holds for most sizes of fish at many operating frequencies. Perhaps the effect of the middle zone in frequency has been a little over-estimated and the variances in target strength measurement should be investigated. Indeed where the statistical nature of the quantitative use of the echo sounder is examined, the variance of any estimates of target strength will play a very important part in the analysis of signals.

5. *The estimation of volume reverberation*

So far we have considered signals as received from single fish or from fish shoals. Sometimes, the whole sampling volume between two ranges may be filled with animals as for example in a scattering layer. Then the volume reverberation signal is the ratio of intensity scattered back by a unit volume to that of the incident wave. The volume in which reverberation is estimated within one pulse length in range ($c\tau/2$), and in a solid angle, Ω, subtended by a cylinder, with ends normal to the incident wave is given by

$$r^2(c\tau/2)d\Omega$$

where τ is the pulse duration. It can be shown that the reverberation level, RL,

$$RL = 10 \log \left[\frac{I_0}{r^4} . S_v . \frac{c\tau}{2} . 4r^2\right] \qquad (29)$$

where S_v is the ratio of intensity scattered by unit volume to that of the incident wave.

$$RL = S - 20 \log r + S_v + 10 \log V - 2ar/1000 \qquad (30)$$

where $V = (c\tau/2)\psi$; where ψ is the solid angle approximating to Ω in the beam pattern of the transducer in echo.

In other words, reverberation in the volume, V, is determined by the approximated beam width, ψ, range, r, and pulse duration, τ, which is short.

Equation (18) ($EL = S - 2H + T$) includes the one-way propagation loss in transmission and in echo (i.e. $2H = 40 \log r + ar/1000$). The corresponding term in eqn. (30) is $20 \log r$ because the reverberation is returned from the volume insonified rather than from a point in it so the volume reverberation is a squared function as compared with echo level. The difference in use is really in analyzing target strength and scattering layers; the theory of volume reverberation requires that the sound is received from all the surface insonified by the solid angle ψ. The theory of echo level requires that there is one target at a particular point in the sound beam. Echo levels can be used for the segregation of sizes in received signals but reverberation level is used for estimates of biomass in a scattering layer, with no reference to the sizes of the animals.

6. *The resonance of air bubbles*

The importance of the swim bladder, as a component of target strength, has been stressed; a rough rule is that half the signal is received from an organ occupying 5% of the fish's volume (in sea water). Hence signals are received from fish without swim bladders, e.g., mackerel, most tuna, sharks, and flatfish. It is well known that a free gas bubble resonates to the transmitted sound at a particular frequency; an increase of up to seventy times in received amplitude is expected and the bubble oscillation distorts the local sound field in such a way that more of the incident energy is intercepted. But the swim bladders of most fish do not resonate at the frequencies employed in most echo sounders. However, it is likely that those of larval fish and small fish do so and the problem of resonance is of great importance in the study of the Deep Scattering Layer in the deep ocean which is probably mainly composed of small fish like myctophids. A later chapter is devoted to the biological and acoustic study of this layer.

If the swim bladder is regarded as a spherical bubble, its scattering cross section should be considered in three parts; at low frequencies the Rayleigh scattering law is obeyed (i.e., $\propto 1/\lambda^{-4}$; 12 dB/octave); at a somewhat higher frequency, the bubble resonates and its scattering cross section is equal to the geometric cross section and so is independent of frequency. For fish of commercial sizes, 20–150 cm with swim bladders, the resonant frequencies range from about 200–1200 Hz at a depth of 50 fm. Because swim bladders are affected by pressure, any comparison of resonances must be made at constant depth.

Weston (1967) has considered the problem of the resonance of fish swim bladders in some detail. The frequency of resonance of an ideal gas bubble, f_0, is given by

$$f_0 = (2\pi a)^{-1} (3\gamma P_0/\rho)^{\frac{1}{2}} \tag{31}$$

where a is the radius of the bubble; γ is the ratio of gaseous specific heats; P_0 is the hydrostatic pressure in atmospheres; ρ is the density of the water.

An expression for scattering cross section was developed; at low frequencies, $\sigma = 4\pi a^2 (f/f_0)^4$, the Rayleigh law of scattering is followed; at resonance, $\sigma = \lambda^2/\pi$ and so is independent of bubble size and at quite high frequencies $\sigma = 4\pi a^2$, so it varies with bubble surface area, and is independent of frequency. Expressions were developed for deformed bubbles, for example, prolate spheroid, which is the shape of a swim bladder.

Swim bladders differ from ideal gas bubbles; thermal effects reduce the resonant frequency slightly and the Q factor, or spread of resonance, considerably. The wall of the bladder and the surrounding tissue increase the spring factor and hence the ambient pressure and the resonant frequency will rise; because fish tissue is more "lossy" than sea water, Q is reduced.

A calculated scattering cross section is given by:

$$\sigma_s = 4\pi a^2 \left[(1 - f_0^2/f^2)^2 + 1/Q^2 \right]^{-1} \tag{32}$$

where $Q^{-1} = Q_r^{-1} + Q_t^{-1} + Q_f^{-1}$. Here Q_r is radiation damping, Q_t is thermal damping, and Q_f is the fish tissue damping. The effect of the damping terms is to reduce and broaden the resonance. Andreeva (1965) has reported some measurements of the shear modulus of fish flesh made by Lebedeva (1965), and suggests that it is important below 200 m. From the equations developed by Andreeva, σ_s increases with depth to a maximum between 150 and 400 m (depending a little on the frequency used, 1–50 kHz), and below 400 m σ_s decreases. Thus, as a fish migrates towards the surface, the scattering cross section near resonance would be expected to fall by as much as a factor of six; at depths greater than 500 m the reverse would be true. So the echo sounders were vulnerable to the detection of the Deep Scattering Layer of small fish which make a daily vertical migration within such a range of depths.

Weston has developed a simplified equation relating resonance to the length of the fish, allowing 6% for an enhancement due to shape distortion:

$$f_0 L = 8P_0^{\frac{1}{2}} \tag{33}$$

The following table gives the resonant frequencies of the swim bladders, of 0.5–100.0 cm in length, at depths from 10–100 m.

TABLE 6. RESONANT FREQUENCIES IN kHz OF FISH OF DIFFERENT
SIZES AT DIFFERENT DEPTHS

Depth (m)	10	30	50	100
Length of fish (cm)				
0.5	16.0	22.4	35.5	49.6
1.0	8.0	11.2	17.6	24.8
10.0	0.8	1.12	1.76	2.48
30.0	0.27	0.37	0.58	0.83
50.0	0.16	0.22	0.35	0.49
100.0	0.08	0.11	0.18	0.25

Thus, in general, the swim bladders of commercial fish resonate at 80–2000 Hz, whereas those of fish in the Deep Scattering Layer do so at 1–15 kHz and those of fish larvae at 10–40 kHz. Swim bladders constitute 5% of the fish's volume; the diameter of a spherical swim bladder would be given by $a = (0.038 \, kL^3/\pi)^{\frac{1}{3}}$, where L is the length of the fish and k is a coefficient of proportionality of volume to length. But swim bladders are usually elongate, perhaps cylindrical or ellipsoidal; they may be compartmented and in some fish, there are capsules extending forward into the inner ear (see Harden Jones and Marshall, 1953, for a full description of the form of the swim bladder in different fishes). Dr McCartney has pointed out to me that shape seriously affects Q, but that the resonant frequency is more a function of volume.

Above resonance, σ varies with the surface area of the bladder; Weston suggests an enhancement of 75% for factors of shape. Then $\sigma = 4.10^{-2}L^2$ and target strength, $T = \sigma/4\pi = 20 \log L - 25$ dB. The most interesting point is that from resonance to $c/L = f$ (where c is speed of sound in cm sec^{-1} and L is length of fish in cm), σ is independent of frequency. In the following table is given the frequency at which $c/L = f$ for different lengths of fish:

L (cm)	Frequency (kHz)
5	30.0
25	6.0
50	3.0
75	2.0
100	1.5

For echo sounding on commercial fish, fish tissue is of importance, and σ may increase in frequency with the area of both bladder and of body. For smaller fish in the Deep Scattering Layer, between resonance and c/L, σ for the bladder increases with area, independently of frequency, but σ for the body increases by the Rayleigh scattering law. So at frequencies below resonance, it is likely that only the swim bladder is of importance.

McCartney and Stubbs (1971) have measured the dorsal target strengths of live fish and swim bladder resonance at the same time with different techniques for different frequency ranges. Cod, ling, pollack, and coalfish were caught by handline, kept in a keep net at 30 m, and were positioned by divers. A wide band CW sound source was used and it was monitored by a hydrophone 1 m from the fish; measurements were made with a ring hydrophone very close to the fish, surrounding it. Measurements were made with and without the fish.

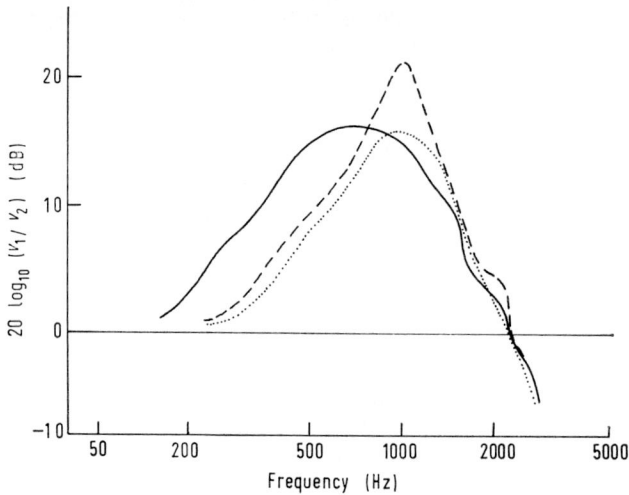

FIG. 47. The resonant response with frequency (McCartney and Stubbs, 1971).

Figure 47 shows the distribution of resonance in frequency in the region of 1 kHz. The resonant frequency is 40–100% greater than that expected from a spherical bubble. The swim bladder is elongated, but the distortion from the spherical only accounts for 20–30% of the increase in frequency. The thick bladder wall and the gut generate a high damping of resonance.

Thus for fish of commercial sizes, the frequencies employed in most echo sounders and sonar equipments are outside the resonant ranges of air bladder sizes. Further, the variability of the resonant response at low frequencies is so great that a quantitative assessment of fish seems improbable; however, in later chapters an account will be given of a method of assessing numbers of fish with low frequencies.

7. Calibration of equipment

Any quantitative work with echo sounders presupposes that the transducers remain calibrated. Many commercial echo sounders are not explicitly calibrated because the display used is essentially an indication of the presence or absence of signal from fish together with a continuous signal from the seabed. The following account is taken from Mitson (1969).

To calibrate a transducer, three items of ancillary equipment are needed, a calibrated test hydrophone, a signal generator (with pulsed and continuous wave outputs), and an oscilloscope. The test hydrophone is lowered on a rigid frame suspended below the ship so that it faces the ship's transducer to be calibrated. Observing the display of transmission, the time elapsed in range and the signal received by the test hydrophone on the oscilloscope, the test hydrophone is moved until the maximum signal is observed at the center of the beam. The distance apart of transducer and test hydrophone is measured on the oscilloscope, given a velocity of sound (which is usually taken to be about 1500 m/sec^{-1} in the sea). Measurements can thus be made from the following equations:

$$S = V_R - S_{Rt} + 20 \log_{10} r \tag{34}$$

where S_{Rt} is sensitivity of test hydrophone in dB V^{-1} mcbar^{-1}; V_R is output across test hydrophone in dB V^{-1}; r is distance between transducer and hydrophone in m.

To calibrate the receiving system, instead of transmitting through the ship's transducer to the test hydrophone, the latter is fired by the signal generator and the signal is received through the ship's transducer. Then,

$$S_R = V_s + S_{tt} - 20 \log r \text{ dB}^{-1} \text{ mcbar}^{-1} \tag{35}$$

(referred to a standard output from the generator) where S_R is the sensitivity of the receiver system; V_s is the r.m.s. output from the signal generator in dB V^{-1} plus receiver amplifier gain setting (dB); S_{tt} is the sensitivity of the test hydrophone as a transmitter in dB mcbar^{-1} V^{-1} m^{-1}; and r is the distance between transducer and hydrophone.

The target strength of a common 8-in. trawl float is known (-23.8 dB) and can be used as a quick check of the acoustic output employing the sonar equation by hanging it below the transducer in the center of the beam. Similarly, a float can be used as a measure of signal strength relative to an unknown target, at the same position, anywhere in the sound field. Recently, my colleague, R. B. Mitson, has devised a means of attaching a semi-rigid structure to a towed body so that calibrated hydrophones can be used at sea as frequently as required.

CHAPTER 4

THE ESTIMATION OF
ABUNDANCE WITH AN ECHO SOUNDER

1. *Introduction*

BECAUSE signals are presented on a paper record with limited dynamic range and because the noise levels are often very variable, the echo sounder was not considered in the first place to be a quantitative instrument. It was designed to measure the depth of the sea in a quasi-continuous manner and as a bonus it indicated the presence of fish at certain times. However, the signals can be measured and it is now possible to make absolute estimates of some fish populations. In the light of this experience, methods of estimating relative abundance can be relied on to a greater extent than previously. So we distinguish between methods of estimating relative abundance, for which an ordinary commercial echo sounder may be used, and methods of estimating absolute abundance for which more sophisticated equipment is sometimes needed.

2. *Methods of estimating relative abundance*

Because the early use of echo sounders was essentially a qualitative one, a mid-water trace was taken to indicate the presence of fish. The first fishermen to use such echo sounders were the herring fishermen in the North Sea and off Vancouver, in British Columbia, who shot their drift nets in areas where fish traces were found.

Historically, the most important work was that carried out by the late Skipper Balls, a drifter skipper, when fishing for herring in the North Sea during the thirties. His echo sounder (a Marconi 424) had a neon tube time base and he found it possible to make a rough estimate of the density of fish below the drifter. In common with most fishing skippers he kept detailed records for a number of years and he distinguished between shots made by his drifter when an echo sounder was used and shots made without it. Figure 48 shows the results of his work: during a period of years, the drifter shots made with an echo sounder or "echo shots" caught twice as much as the ordinary shots and Skipper Balls' use of the echo sounder increased from about 5% of the time to more than 20%. The important conclusion from this work was that the presence or absence of fish was reliably established and that some estimate of quantity was made (Balls, 1946).

The step to echo survey was taken by Runnstrøm (1941) and Sund (1943) who published surveys of herring off the Norwegian coast. Runnstrøm's survey showed fish traces on the Utsira spawning grounds of the Norwegian herring during February 1937 and the presence or absence of fish is shown in the area surveyed (Fig. 49). Sund (1943) published distributions of fish echoes along the whole length of the Norwegian coast, which at first sight seemed unwarrantably ambitious. But in the early forties, the stock of Atlanto-Scandian herring was very large and from later experience of catches of the purse-seine fleets, it is

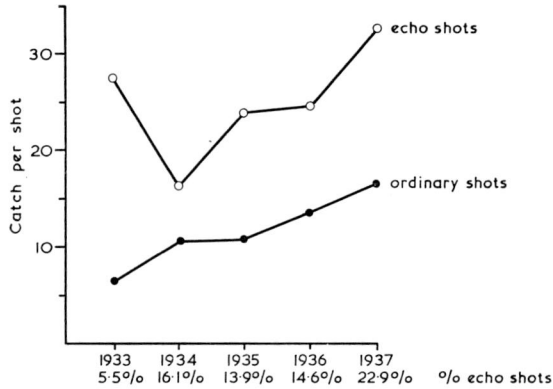

FIG. 48. The results of the pioneer work of the late Skipper Balls, in showing that echo shots yielded better catches than blind shots; the figure also shows greater reliance on echo shots, as time went by (Balls, 1946) [reproduced by courtesy of the Marconi International Marine Company Ltd.].

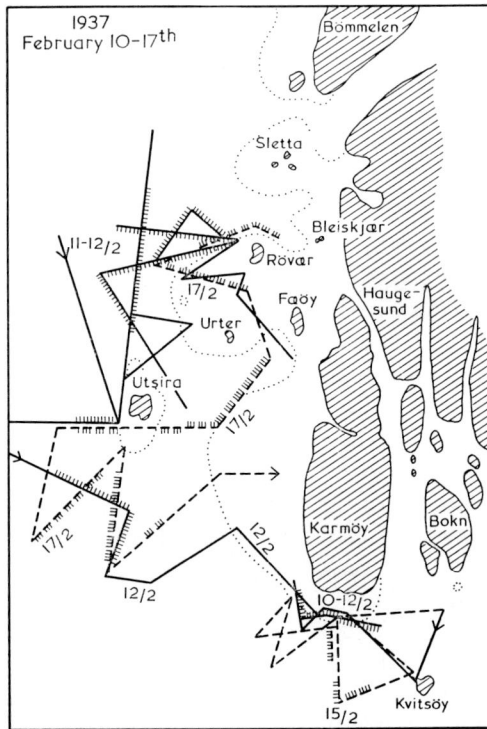

FIG. 49. The first echo survey made on the Utsira spawning grounds of the Norwegian herring (Runnström, 1941).

unlikely that other pelagic fish, sprats, mackerel, or capelin, were as abundant as herring off the Norwegian coast.

During this period, the fish were recorded mainly as shoals and, as they lived roughly at the same depth, the physical requirements determining presence or absence of fish were fulfilled. That is, in the same depths of water, the noise levels, which are mainly due to propeller noise, were roughly constant (for a given wind strength and sea state) and so if fish shoals were detected in one part of an area, their apparent absence in another part indicated true absence. Later, when it became possible to distinguish individual fish on the paper record, the absence of signal was not so reliable an index of the absence of fish, because the signal to noise ratio was less; with weak echo sounders (as compared with those available today) signals from individual fish were much more vulnerable to noise variations than those from fish shoals. With constant gain control on its echo sounder, a survey ship might steam into shallow water and individual fish signals properly recorded at the same depth in deep water would disappear into noise merely because propeller noise increases in shallow water. If the gain was altered to take this factor into account, the system might yield uncontrolled results, although, in later years, deliberate gain changes were used as a primitive form of time-varied gain. However, in the early surveys this problem did not arise because the signals from fish shoals were very large, or in physical terms, the signal-to-noise ratio was very high.

Tester (1943a, 1944) made preliminary surveys for herring in the Strait of Georgia between Vancouver Island and the mainland off British Columbia in western Canada and Smith (1947) described distributions of sardines off California in the U.S.A. Tester's work on Pacific herring was continued by Hourston (1953). In the North Sea, surveys for herring were made in limited areas by von Brandt and Schärfe (1950), Krefft and Schubert (1950), Le Gall (1952), Craig (1952, 1953, 1954), Parrish (1952), and Percier (1953). Sprats were surveyed with echo sounders in the Thames estuary on the edge of the southern North Sea by Richardson (1950, 1951) and the distribution of sardines off Morocco was described by Furnestin (1953). Chapman (1944) and Chestnut (1950) used recording echo sounders to chart seaweed beds. All these echo surveys were made in the same way as Runnstrøm's first one, by marking positions where fish were recorded on an Admiralty chart. Most of the surveys listed above revealed the presence of fish in particular areas, but there was little attempt to relate the information gained to fish catches or to environmental characteristics; the scientists concerned were really convincing themselves that an echo survey method was a practicable proposition. That made by Schüler and Krefft (1951) was a rather more developed form of survey in that the horizontal and vertical extent of each trace was measured. This was a considerable advance, but it was sensitive to the effects of reverberation described in Chapter 2 and so may have over-estimated the relative densities to some extent. However, such an effect would have been unimportant because the differences in abundance between stations were very much greater than the small over-estimates due to reverberation.

Cushing (1952) extended this method and obtained an estimate of the horizontal extent of the trace by measuring, with a binocular microscope or a ruler, the width of the mark made by successive transmissions on the paper record. Thus each transmission was explicitly examined for the presence or absence of echoes in mid-water. Because the signal-to-noise ratio used on the early echo sounders tended to be low, and they were vulnerable to various forms of interference, a signal from fish was defined as one repeated at about the same depth in successive transmissions. In other words, a form of trace-to-trace correlation was used and, while the method cannot be used to determine the absolute abundance of fish directly, it can, and has, been used to give estimates of relative abundance. It was applied in a number of investigations, especially of pelagic fish (Cushing, 1952). Figure 9 shows a relationship between the distribution of pelagic fish and the distribution of *Calanus*, the preferred food of the herring: patches of fish coincided with those of *Calanus* and this method was used to estimate the aggregation and disengagement of herring (and other pelagic fish) on patches of their food off the north east coast of England. In the course of his studies of the North Icelandic herring fishery and the association of herring traces with patches of *Calanus*, Jakobsson (1963) has shown that such processes occur repeatedly on a rather larger scale. Both systems were described in detail in Chapter 1 (Fig. 10).

Figure 50 shows an echo survey of pelagic fish in the English Channel in the summer of 1950. At the same time samples of pilchard eggs were being taken from R.V. Sir Lancelot (Cushing, 1957). There is an association between the two distributions, of eggs and echo traces, which provides an identification for the great majority of echo traces in the area. As estimate was made of the number of fish in a shoal recorded as a daylight plume; the number of "plume" traces on a cruise was divided by the number of pilchards, as estimated from the distribution of pilchard eggs, to give an average value of 4000 fish per shoal. Because a small proportion of plumes was probably not composed of pilchards, this average density within a shoal was under-estimated. Until the work of Truskanov and Scherbino (1964), on the direct estimates of numbers within a shoal, this was the only estimate available.

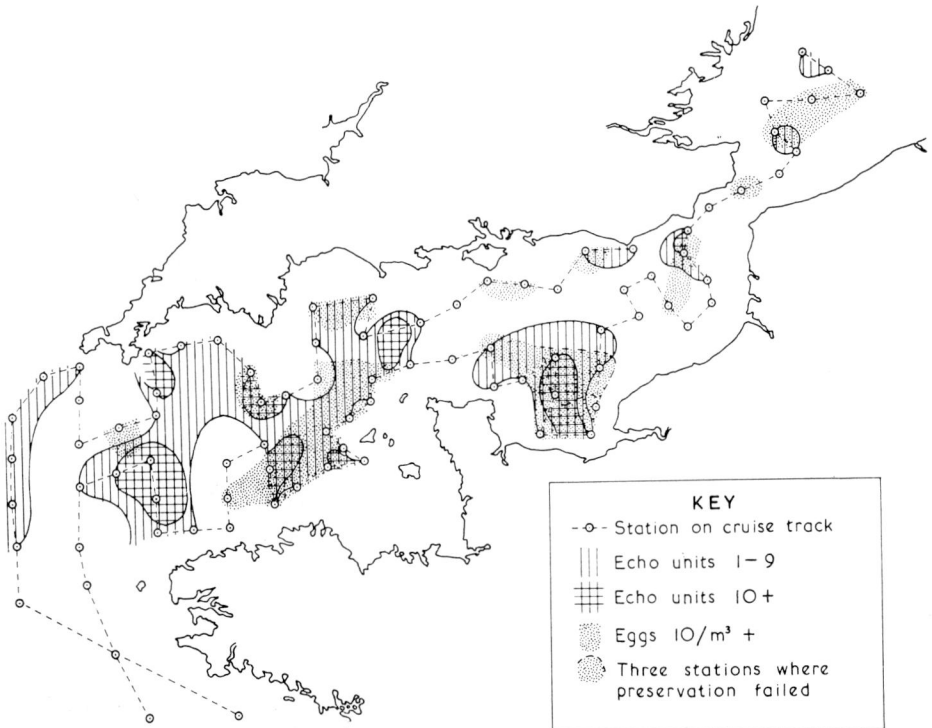

KEY
- -o- Station on cruise track
- ||| Echo units 1 − 9
- ‡‡‡ Echo units 10 +
- Eggs 10/m³ +
- Three stations where preservation failed

FIG. 50. An echo survey of pelagic fish in June 1950, in the English Channel, associated with distributions of pilchard eggs observed at the same time (Cushing, 1952).

Such methods of estimating relative abundance were developed in the early fifties and are used at the present time. A long series of such surveys was made in the Southern Bight of the North Sea during the late fifties to assist the East Anglian driftermen when their fishery for herring was collapsing (Tungate, 1958); Fig. 51 shows a typical survey in mm trace per

10 miles steamed (Valdez and Cushing, 1966). For three years, 1956, 1957 and 1958, the results of all cruises were plotted on one chart, showing that echo traces were concentrated on the eastern edges of the Norfolk Banks, on the Brown Ridges, on the Schouwen ground, and on the herring spawning grounds in the south. These grounds were familiar to the driftermen, and may be the concentration area on the southward migration of the herring. The three regions are areas at which the driftermen gathered; if the fishery failed on the Norfolk Banks, the skippers would steam directly to the Brown Ridges or the Schouwen ground. They were the traditional grounds which were discovered long before echo sounders were introduced.

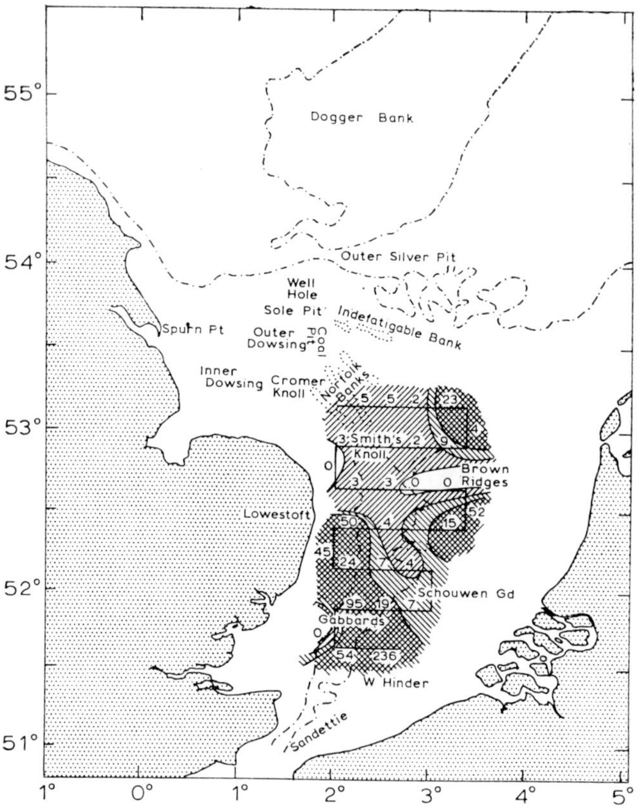

FIG. 51. An echo survey of pelagic fish charted in mm per mile during the autumn in the southern North Sea, when the Downs herring were migrating south to their spawning grounds near the Straits of Dover (Valdez and Cushing, 1966).

Combined surveys with three or more ships from the three countries in Scandinavia were used to chart the oceanic distribution of the Norwegian herring in summertime (see Fig. 4). More recently, ships from Norway, U.S.S.R. and the U.K. have surveyed the surface waters of the Barents Sea (Dragesund, 1970). The scattering layers at about 30 m were charted and identified by capture. In this way, distributions of the small fish (less than a year old) have been described over the whole Barents Sea for a number of years; the species were herring, cod, haddock, capelin, saithe, redfish, polar cod, and long rough dab. For four years 1965–8, indices of recruitment for nine species were given. Such surveys of are considerable biological importance during a period when the Arcto-Norwegian cod stock suffered much pressure upon its recruitment. An interesting point is that voltages have been summed on an integrator (Dragesund and Olsen, 1965), which is a first step towards the automatic processing of data. The method will be examined again later in this chapter.

A considerable series of echo surveys (Fig. 52) has been executed in the anchoveta fishery off Peru as described in Chapter 1. Figure 24 shows a zooplankton survey associated with a distribution of anchoveta eggs. As the latter comprise 90% of all the eggs caught (Flores *et al.*, 1967), it is concluded that most of the echo traces were in fact anchoveta traces. Figure 52b shows a vertical section showing the echo traces in the upper part of the thermocline. The surveys are particularly valuable because the fish traces represent the dominant pelagic fish species and the dynamics of the whole simple food chain could be analyzed with echo survey playing a prominent part in the method.

A particular form of survey known as the "Eureka cruise" has been developed off Peru. A number of fishing boats were chartered and each steamed to sea from a well marked navigational point; after steaming some thirty miles, the boat returned at an angle to its original course, making landfall at a second well marked navigational point. So each boat executed a dogleg with its point to seaward. The survey consisted of ten to thirty such doglegs set alongside each other and the whole coast was surveyed in a single night. Figure 52 shows an example of such a survey. The method was devised by Vasco Valdez and gives immediate results for the whole fishery area; if a single boat covered the area, the information from the early part of the survey might be stale by the time the survey was completed. If the stock were to be examined in great detail in time, Eureka surveys executed every night for a period of 10 days would yield the most valuable information on the movement and on the stability of patches of anchoveta.

In recent years the tendency has been to use the methods of estimating relative abundance in rather more sophisticated ways. For example in a Special Fund project of the U.N., (Report 69/1) off West Africa a considerable effort is being made to survey the resources in the area with echo sounders. The same Simrad machines are used in five countries and they are all calibrated. The most important point is that a standard log sheet is used on which is recorded the number and extent of the fish schools, the extent and character of scattering layers, and the numbers of single fish. Four categories of shoal were used, assuming that shoals were vertical cylinders, in relative volumes of 1, 4, 16 and 64, respectively. A correction for beam width had to be made. The four categories, before correction, were 50, 77, 121 and 193 m in horizontal extent, respectively and 8, 12, 18 and 27 m in vertical extent, respectively. The assumption that each shoal is a vertical cylinder will under-estimate the true volume of a shoal. Olsen (1969) shows that the volume of a truly cylindrical shoal is under-estimated by about 40%. For ellipsoidal shoals, the bias is variable and may even become an over-estimate. If small shoals are cylindrical and large shoals ellipsoidal, then the small ones are under-estimated and the large ones are over-estimated. This method has been treated in a little detail because it is based on a quick method of shoal classification and, from a sum of the conventional categories, shoal volumes can be estimated. Considerable quantities of information can be handled in this way. Analogous procedures are being used by Bullis at the Bureau of Commercial Fisheries Exploratory Fishing Base at Pascagoula, in the U.S.A. in the Gulf of Mexico; forms of log sheet are used on which diurnal, seasonal, and depth variation of a few selected types of echo trace are tabulated. The classification of records over a period of survey is a most important part of the development of

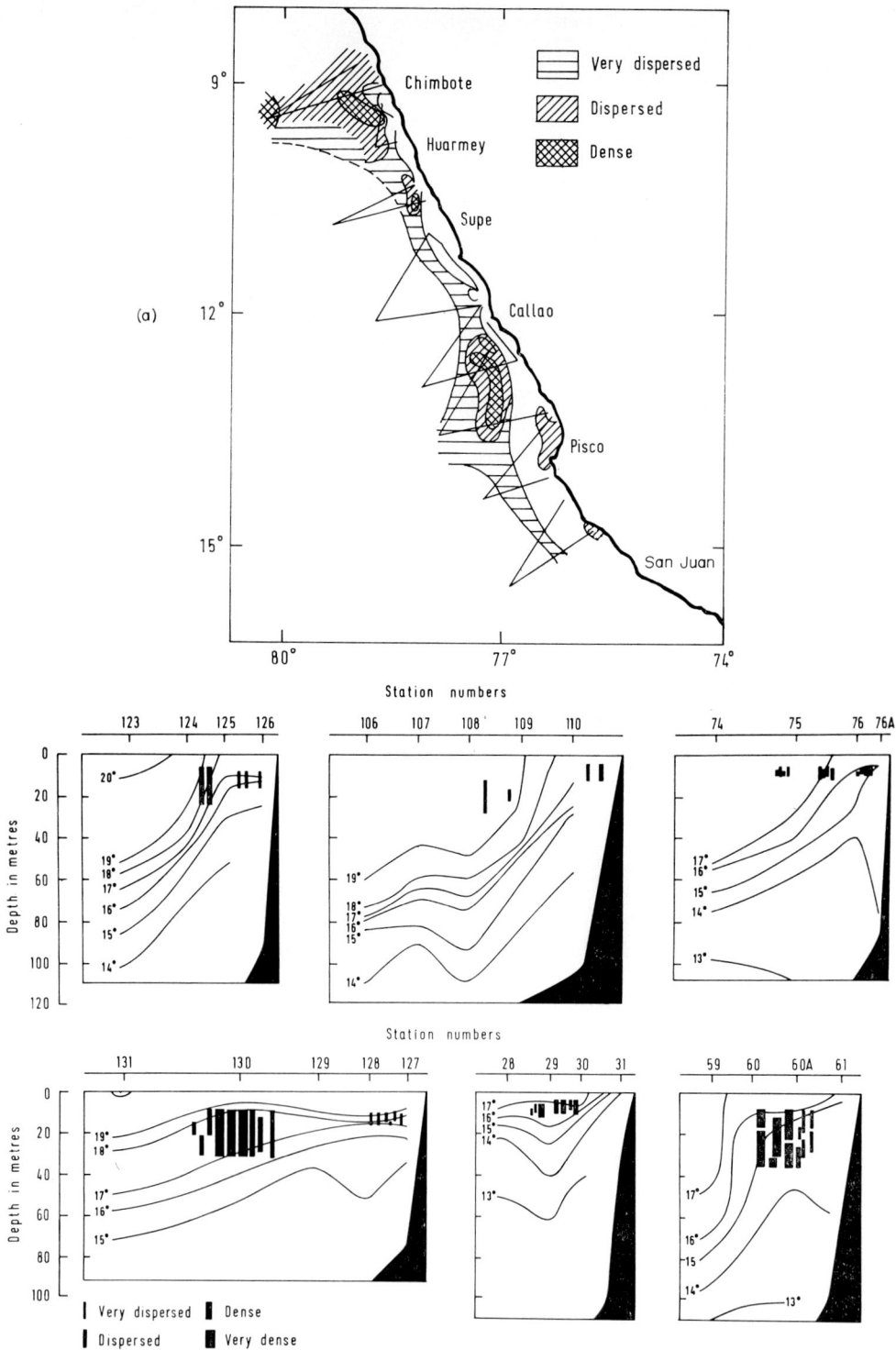

FIG. 52. a, A survey made by a Eureka cruise, made by a number of fishing vessels on a single night (Flores and Elias, 1967). b, The vertical distribution of echo traces in the area of the Eureka cruises (Flores and Elias, 1967).

resource exploration, because the accessibility and vulnerability of fish stocks at different seasons and times of day varies greatly.

Alverson (1967) has surveyed the hake stocks off Oregon and Washington State in the U.S.A. with echo sounders and mid-water trawls. Two stocks have been discovered, that which returns to the north west Pacific each summer after spawning off California in winter and early spring and that which remains off Port Susan in Puget Sound all the winter. Because these were exploratory fishing surveys in which mid-water trawl hauls were frequently made, Alverson was able to make immediate estimates of the offshore hake stocks in summer time and to make forecasts about their subsequent exploitation. The survey which accompanied the mid-water trawl exploration was essentially one for estimating relative abundance: the estimate of absolute abundance was derived from the trawl hauls made in the areas in which fish were detected with the echo sounder, assuming that a trawl caught the fish in its path.

In conclusion, the method of estimating abundance in a relative manner with an echo sounder has been of great use if handled carefully. Today we would say that the sounder should always be calibrated and it is likely that differences in abundance between stations will be much greater than differences due to physical factors. The diverse information should be categorized and carefully logged to take account of seasonal, diurnal, and depth variations. The survey grids should be carefully arranged to circumscribe the biological area under investigation and, if possible, they should be repeated frequently. Lastly, the ancillary physical and biological data which may be collected on a cruise are often of great value in interpreting the distribution of echo traces, for example the vertical distribution of anchoveta as echo traces in the thermocline off Peru and the horizontal distribution of echo traces of anchoveta, and the distributions of anchoveta eggs, guano birds, and zooplankton in the same area.

3. The use of sonar to estimate the relative abundance of fish

Because sonar had been used in World War II to detect submarines, attempts were made immediately afterwards to use it to find fish. Smith (1947) and Smith and Ahlstrom (1948) published observations, on sardines, off California. Renou and Tchernia (1947) surveyed the winter herring fisheries in the eastern English Channel with sonar and subsequent work showed that the patches of fish which they located were in fact the spawning grounds of the Downs herring (Cushing, 1966). But the most dramatic work with sonar was that started by Devold (1951, 1952) on the migration of the Norwegian herring towards the coast of Norway.

The Norwegian herring fishery used to work on the spawning shoals between Utsira and Bergen (see Chapter 1). A great tagging experiment (Fridriksson and Aasen, 1952) executed partly on the spawning shoals and partly on the summer feeding shoals off north west Iceland had shown that the herring migrated between the two areas. Hence to chart the migration routes of the herring in the Norwegian Sea, Devold made sonar surveys in the Norwegian Sea, supposing that the fish had migrated from Iceland in the East Icelandic current towards the Faroe Islands and Shetland Islands. Figures 3, 4a and 4b in Chapter 1 describe the nature and the scale of this work; not only was the arrival of fish in Norwegian coastal waters predicted for the fishermen, but a new autumn and winter fishery eventually developed in the East Icelandic current. For the fisheries biologist a reliable picture of the migration circuit of the Norwegian herring was evolved.

Jakobsson (1963) has executed similar surveys in the area north east of Iceland and has shown the movement of the herring southward from the polar front and on to the *Calanus* patches as described in Chapter 1. Later, the movement of the fish southwards in the East Icelandic current was established. Part of this work has a biological basis as described, but part was a scouting survey for the Icelandic fleet, which in recent years of herring lack, has ranged as far afield as the Barents Sea. The scouting survey shows where the general area of abundance is, but also can indicate the positions of particular shoals, which would merit attack by purse-seiners.

Smith (1970) has succeeded in using a Simrad sonar which transmits at 90° to the ship's track, in other words, as a rather wide angle side scanner. It was used at an intermediate range and errors due to overlap were taken into account. A chart of the horizontal distribution of fish shoals (anchovies and sardines) was prepared and, by comparing catches and shoal distributions, Smith was able to make an estimate of the biomass of pelagic fish in an extensive area off California. The oceanic structure was variable because there were internal waves in the thermocline; but, the presence of fish shoals was charted, as above or in the upper part of the thermocline. It remains possible that the positions of some shoals relative to the ship's track might have been plotted inaccurately because the beam was distorted, but this does not matter.

Sonar is used to estimate the relative abundance of fish *shoals* in an area by their presence and absence and has been particularly useful in the north east Atlantic. This is because the herring live above the seasonal thermoclines and so acoustic conditions are roughly the same over the whole area and in all seasons. Where fish live in the thermocline, as does the anchoveta off Peru, detection might be possible on some occasions, but it would be rather unreliable. The quantitative use of sonar is practically impossible for two reasons, (a) because of surface reverberation, intensity does not decay predictably as the one way propagation loss, but decays approximately as the inverse cube and in a variable manner and (b) because a transmission at any angle to the surface must be refracted by any density layering in the sea, the range of a good target like a submarine can only be determined accurately if the degree of refraction is known, according to the distribution of temperature and salinity with depth. At any real angle of tilt the problems become complex because the rays may be skipping, bouncing, or just disappearing. For a target like a fish shoal which is diffuse and variable in intensity (by the attitude of the fish which vary from side aspect to head aspect), the problems are greater. This is not to say that fish shoals cannot be evaluated, because Icelandic purse-seine skippers, in particular, use sonar with very great skill and success, in finding shoals, following them and capturing them; indeed they can predict the quantities which they are about to catch. This implies that the quantity is a direct function of the apparent volume (as a sonar trace) and that the packing density of the herring must remain constant. It is an example of where the fishermen's practice has run ahead of the scientists. It will be shown below that the quantitative use of the echo sounder depends upon the precise evaluation of target range, in a statistical manner. This step, for the reasons given above, must remain inaccessible to sonar practice in general and it is only in this sense that sonar is considered to be limited to the estimation of relative abundance. When the fisheries biologists come to rationalize the techniques exploited by the Icelandic purse-seiners, an estimate of absolute abundance with sonar will be developed.

4. *The detection of individual fish*

The first steps towards establishing surveys of relative abundance were taken in the

size of fish can be determined in terms of the sonar equation. Then it is only a short step from the survey of individual fish to the estimation of absolute abundance, in expressing the numbers as densities when the methods of estimating sampling volume are fully developed.

5. *The estimation of absolute abundance*

The basis of the estimation of absolute abundance lies in one of the sonar equations, (eqn. (18) in Chapter 3):

$$EL = S + T - 2H$$

From the source level, S, one way propagation loss, H, and the target strength, T, the echo level, EL, at the maximum range can be calculated, given a conventional signal-to-noise ratio. There are three steps in the determination of absolute abundance: (a) the estimation of noise level, (b) an examination of the statistics of signals received in angle, and (c) the determination of the volume sampled by an echo sounder.

(a) *Estimation of noise and the determination of maximum range*

There are many sources of noise in the system as indicated briefly in Chapter 3. There is instrumental noise which may limit the whole system and there are sources of interference, which should be eliminated. Aboard ship, there are often many motors outside the engine room, which may generate interference. Cables are often laid in ships to the convenience of the builders which may create fields to which the receiver of the echo sounder is sensitive. All such sources of interference must be eliminated, not only because the detection capacity of the equipment is reduced by an unnecessarily high noise level, but because automatic counting equipment may well record interferences as signals from fish.

There are sources of noise in the sea itself. When the receiver is switched on, with no transmission, a noise is detected, due to several sources; the flow of water across the face of the transducer, the noises of surface waves echoing into the ocean, oceanic turbulences, tidal scours, and the noises of animals. The receiver does not distinguish between the different sources of noise, but some behave differently from others. For example, as the wind increases so does the surface noise, eventually to make detection impossible; at the same time, the ship creates noise as it surges and slams in the sea and it sweeps bubbles across the face of the transducer and they may obliterate the signal. The bubble layer can be avoided by towing the transducer on a short cable, which also reduces the noise radiated to the transducer from the ship's engines and propellers. Of course, the tidal scours from the gravels and rough sands and the noises from animals are restricted to particular well-defined areas. Indeed the snapping shrimps were discovered in World War II by the hydrophones of submarines entering and leaving Chesapeake Bay on the eastern seaboard of the U.S. The most important single source of noise in echo sounding is that generated by the ship's propeller. To some extent a transducer well forward, or towed on a cable, is shaded from the propeller, but the sound is bounced back from the seabed directly into the face of the transducer. So the propeller noise increases in shallow water. Thus two general principles emerge, that noise in the receiving transducer increases with increasing weather and that propeller noise decreases with increasing depth. Ambient noise decreases with frequency, to about 100 kHz, at which the limiting noise is the thermal noise in the water. Consequently, in designing an echo sounder for a particular purpose, a sonar equation is used with the frequency-dependent noise added to obtain a figure of merit for a particular design.

When the transmitter is switched on, an important component of noise is added to the system. The transmitted signal is returned to the receiver by many paths, some of which may be delayed considerably and therefore reduced in intensity. Thus the reverberation decays exponentially from a given transmission and with a steady pulse repetition frequency, takes up a steady level. However, there are exceptions: the noise increases as the echoes take time to bounce about the rough ground or on a steep slope they return from a shorter path length in shallower water.

Noise is additive and the different sources cannot be distinguished at the receiver and so the variation in noise appears to be capricious. Hence the noise levels should be measured at frequent intervals as changes in depth and in weather take place. Ideally a continuous record of noise should be available if the equipment is to be exploited to the fullest advantage. If this is impossible, then the signal-to-noise ratio should be set at such a level as to take the probable variations into account. Then the maximum range is safely determined and the whole system can be used up to that range.

At the maximum range, the target is detected on the acoustic axis. Then when the noise has been measured, the echo level (EL) is set at three or five times the noise level. From the sonar equation, the range at which this echo level is found can be calculated for a fish of a given target strength. It follows that fish of the size corresponding to that target strength can be detected to the range at which the conventional echo level is found and this is the maximum range, r_{max}. Hence there is a maximum range for each size of fish, increasing with size. It follows that up to a maximum range for a small fish, all larger fish can be detected. So, in an echo survey of absolute abundance, the range of sizes of fish expected should be specified before it starts. It will be shown below that an essential part of the procedure is to make estimates of size of target from the mean amplitudes of the received signals. This depends upon the estimation of target strength as described in Chapter 3. The variability of such estimates is necessarily high, but, with the large quantities of information available, much of the variance can be reduced.

Summarizing, given a measured noise level, the maximum ranges for an array of fish sizes can be calculated. The noise must be monitored continuously during the survey and, if conditions deteriorate, the echo level may rise which means that the maximum range of the system becomes reduced.

(b) *The statistical treatment of signals in angle*

If all fish were found at the axis of the sound beam, differences in signal at any one range would be differences in size. But fish must be randomly distributed with respect to the beam and the signal received within the beam falls away as angle θ from the axis to the first minimum; the squared coefficient is used for signals in echo and so, except at very short ranges, the side lobes can be ignored. Numbers of fish, however, increase as the square of the angle θ, because the beam spreads over an area at any one range. Let the volume insonified be a cone of height ra_1, with base of radius ra^1 and apex angle 2θ. Numbers of fish, n, on the base of the cone increase as the square of the radius, ra^1. Also $ra^1 = ra_1 \tan \theta$ and $\tan \theta = 1/ra_1 \sqrt{n/k}$, where k is the coefficient of fish density. Figure 54 shows the two distributions in angle θ. They can be multiplied together to give a product distribution, the broken line in the figure. The mean of this product distribution in angle is the mean angle, θ_m, at which signals from the fish are most commonly received.

At θ_m the directivity coefficient is $(\delta_m)^2$ and so the signal on the axis is multiplied by $(\delta_m)^2$ to give the expected average signal from the whole beam. The product distribution is extended in angle to θ_{min}, at $0.05\delta^2$; a conventional value is used because at the first minimum itself, $\delta = 0$.

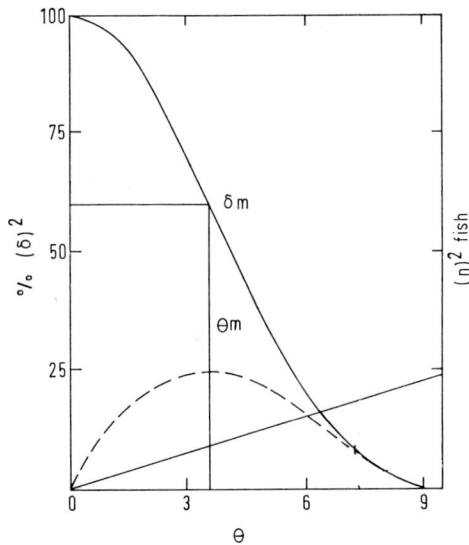

FIG. 54. The distribution of directivity from the acoustic axis in angle, θ, and that of fish numbers in angle, θ, the product distribution peaks at the mean angle, θ_m (Cushing, 1968).

The sampling volume of an echo sounder is described by the maximum range (r_{max}) of the target (of a fish of given size) and the beam angle, i.e., θ_{min}. At a less range r_i, the same signal is recorded at same signal-to-noise ratio at (δ_{min}). Therefore:

$$r_1^{\frac{1}{2}} = r_{max}^{\frac{1}{2}}(\delta_{min})^2 \qquad (36)$$

Between r_1 and r_{max}, θ_m tends towards zero and $(\delta_m)^2$ tends towards unity. At intermediate ranges, taking ratios of r^1/r_{max} where r^1 is any range intermediate between r_1 and r_{max}, intermediate values of θ_m, θ_{min}, $(\delta_m)^2$ and $(\delta_{min})^2$ are calculable.

Thus, the mean angle, θ_m, and the mean directivity coefficient, $(\delta_m)^2$, for a fish of given target strength, vary with range between r_1 and r_{max}. The echo level is calculable:

$$EL_m = EL\,(\delta_m)^2 \qquad (37)$$

where EL is the echo level on the acoustic axis, at a given range, in dB; and EL_m is the echo level at the mean angle in decibels at a given range.

Given the law of target strength on log length, the source level of an echo sounder and the one-way propagation loss, EL_m, is calculable for all ranges to r_{max} for a given size of fish.

Figure 55 shows the calculated relation of echo level in mcV on range for fish of different target strengths; the relationship was calculated for a Kelvin Hughes Humber gear with a source level of 128.9 dB ref. $1\mu b$ and a beam angle of $9° \times 14°$ in echo to the first minimum. An average beam angle of $11.8°$ was used. The beam was supposed to be elliptical in cross section and the average was calculated assuming that the directivity patterns abeam and fore-and-aft were the major and minor axes of an ellipse; J. G. K. Harris in Cushing (1968a) showed that the approximation was precise to within 1%.

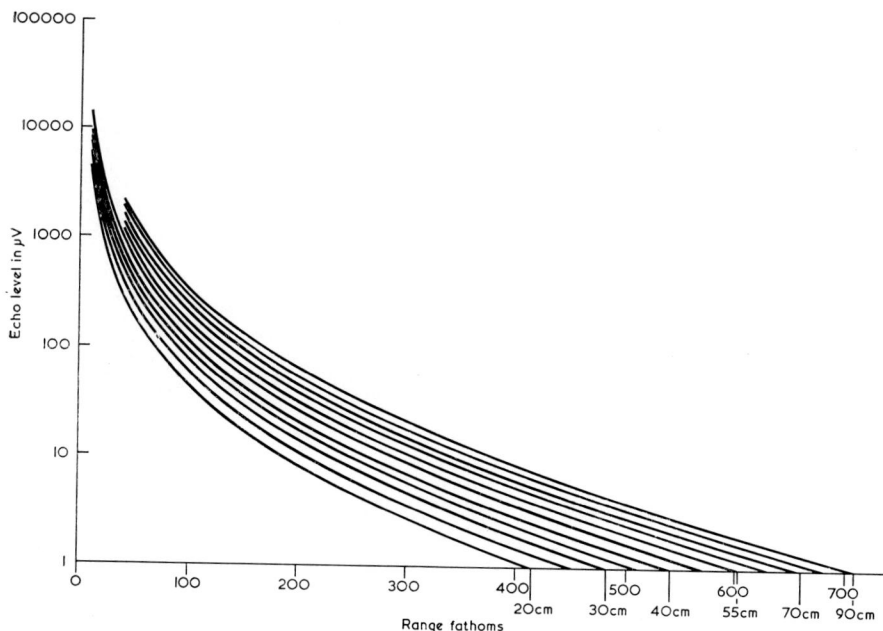

FIG. 55. The dependence of signal expected from fish of different sizes in mcV on range, for a Kelvin Hughes Humber gear (Cushing, 1973).

The same procedure can be used in reverse to make an estimate of the size of fish. At any range, the mean signal, EL_m, from a single fish can be measured in mcV (or in decibels on a calibrated oscilloscope). The signal from a fish on the oscilloscope as it crosses the beam increases in amplitude and shortens in range to maximum amplitude at minimum range and then it decreases in amplitude and increases in range. With practice a mean signal can be estimated in a number of transmissions but is best done with a sampling voltmeter. Then the real problem is to decide how many observations are needed to establish a good average. An average signal strength must be based on measurements made on many signals, enough to randomize the systematic trend and enough to establish the mean. Hence the mean signal, as from θ_m, is averaged from all the signals received from a single fish, or a group of fish; it is not the average maximum signal, the distribution of which is not centered about θ_m. Thus between the transducer and the maximum range for a fish of given size, the mean angle, θ_m, is calculable throughout the sampling volume, the volume within which a target of given size can be detected under given conditions of noise. For each size of fish, there is a specific sampling volume (see below), a maximum range and a mean angle, θ_m, at each range.

Figure 56 shows the calculated lines of EL at θ_m for different sizes of fish on range using a Kelvin Hughes Humber echo sounder. The points represent observed average values of EL at θ_m at different ranges in the 4 fm scan off the seabed using the locked-bottom CRT of the Humber gear. The work was carried out on the hake stock off South Africa aboard H.M.S. Hecla during the month of February 1966. The dotted line shows the trend of average signal with depth showing that bigger fish live in deeper water. The large circle at 210 fm represents the mean size of fish caught by trawl during the period of the survey in

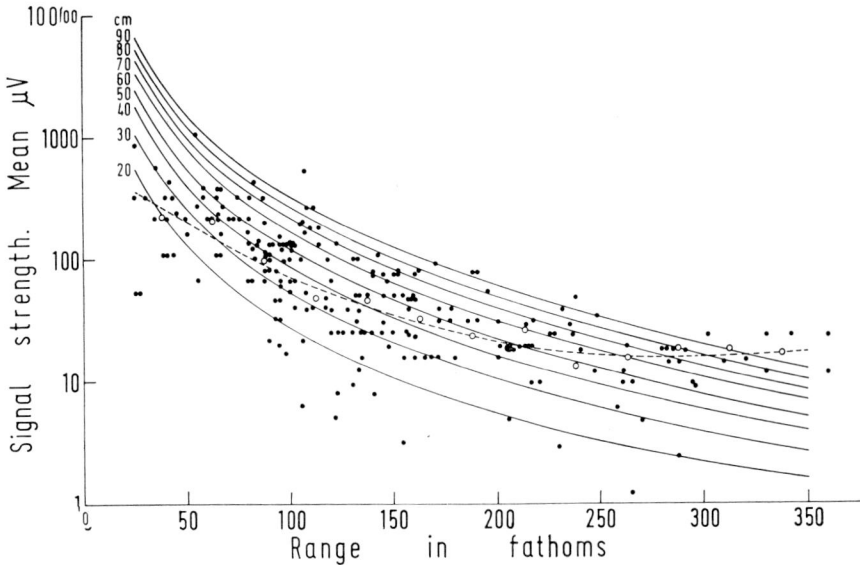

FIG. 56. The calculated dependence of echo level in mcV on range for a number of fish sizes; the points represent the average signals observed for each $\frac{1}{2}$ hr of survey off South Africa; the dotted line represents the mean signal at each interval in depth of 25 fm; the large circle in 210 fm represents the mean size of fish in the catches during the period of survey (Cushing, 1968).

three different areas; it will be seen that the mean size of fish caught is quite close to that estimated acoustically. The range of sizes sampled by trawl and echo sounder were very similar. This is because the oscilloscope was set up to record the amplitude of a 20 cm fish at a low but constant value on the oscilloscope for any depth. As the gain had to be changed with change of depth to achieve this, the gain was effectively calibrated in depth. In other words a primitive form of time-varied gain was being used. The oscilloscope face was graticuled with a scale of 1–10 and so voltages at any depth could be averaged and logged. In the Humber gear system, 1 mcbar = 36 dB above 1 mcV and so the voltage can be converted to decibels in the echo level form of the sonar equation, if so needed.

Figure 57 shows the theoretical curves of size of fish on range calculated in decibels, rather than in mcV; if the signals were calculated as on the acoustic axis, then the curves would be linear, as decibels on logarithmic range. Calculated as at θ_m, the plots remain curved even on a log–log basis. For different sizes of tuna, Nishimura (1963) has plotted "reflection loss" $(T.r^{-2})$ in decibels on log depth; observed signals from fish are scattered about the theoretical lines in much the same sort of distribution as in Fig. 56.

In the sonar equation, the one-way propagation loss accounts for the effects of range and attenuation. The same effect can be achieved with a time-varied gain. When noise is measured, a threshold setting on such a gain is put at a conventional signal-to-noise ratio. The effect of this procedure is to record all signals from fish on the acoustic axis on the same scale and so differences in signal are then only differences in size of fish. However, fish are distributed in angle and at any one depth there are differences in signal due to angle as well as to size. Figure 58 shows the signals expected from different sizes of fish at different ranges with a Humber gear and a calculated time varied gain, at one threshold setting. The most important point about this figure is that the differences in signal for a given size of fish for

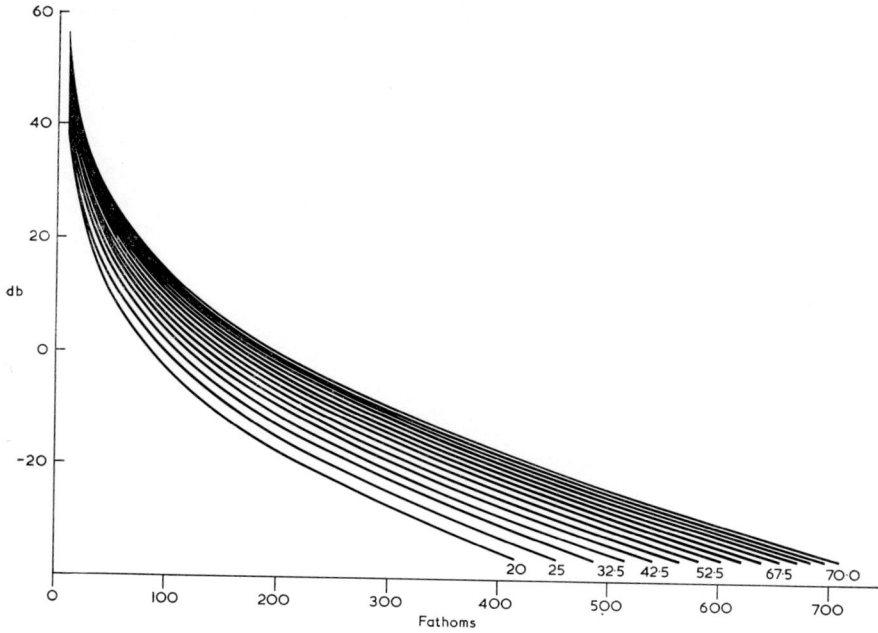

FIG. 57. The dependence of signal expected from fish of different sizes in dB on range, for a Kelvin Hughes Humber gear (Cushing, 1973).

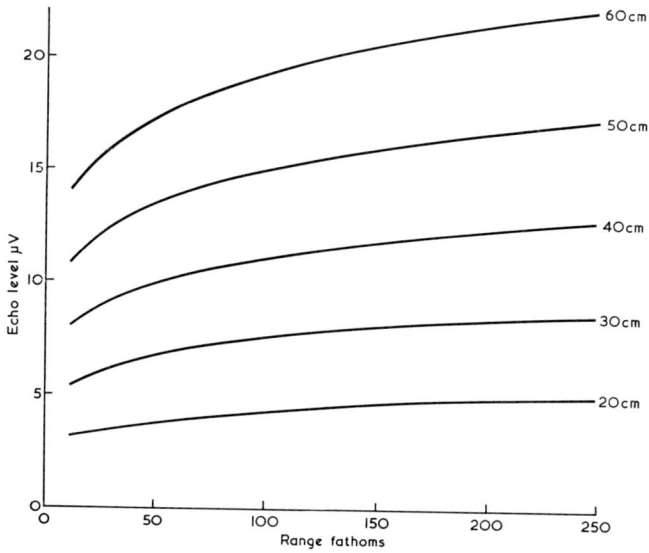

FIG. 58. The signal expected from fish of different sizes in range with a time varied gain, using a Kelvin Hughes Humber gear (Cushing 1973).

a depth range of up to 25 fm are not very great. Consequently, when the noise level is measured, the gain threshold is set at a convenient signal-to-noise ratio and then the system can give signals from fish by size only on the display over quite a convenient range of depths.

Thus, theoretically, it is possible to size fish using the relationships shown in Figs. 56 and 57, and from Fig. 56 there is a little evidence that the method works in practice. It is based on the averaging procedure, by which the mean of all signals from single fish is estimated. If the object is only to size fish from the average signal, then a sampling voltmeter can be used and the period of storage depends upon the rate of sampling. A pulse height analyzer produces a frequency distribution of pulses in voltage. The mean of such a distribution corresponds to a mean size of fish to which a given directivity distribution corresponds. As will be shown below, the Aberdeen method of data processing includes a pulse height analyzer at the present time.

(c) *The mean height above the seabed*

It is also possible to estimate the mean depth (or mean height above the bottom) using the mean angle, (see Fig. 59) given by:

$$h = r_1 - \left(\frac{r_1 + r_2}{2}\right) \cos \theta_m \tag{38}$$

It can be shown that the mean height above the bottom (or the mean depth in a mid-water range slice) estimated in a scan of $(r_1 - r_2)$ varies with the size of fish, because θ_m does so.

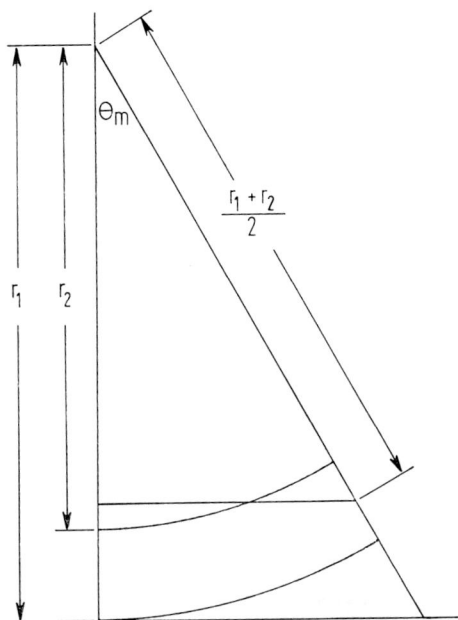

FIG. 59. The mean height of fish above the bottom (Cushing, 1968).

Conversely, if one is interested in the true height above the bottom of, for example, 2 fm, the height of the headline of a trawl, the scan examined $(r_1 - r_2)$ varies with range and size of fish. It increases with range to a maximum and then with a further increase of range the appropriate scan decreases; for different fish sizes the curve is scaled differently. This treatment assumes that fish are randomly distributed within the scan; if they are in a thinner layer or hard down on the bottom, then the scan should be reduced. The important point here is that a proper sampling volume is being used; unless an appropriate sampling volume is used correlations between catch and signal are of little value. Consequently, a counting system for correlating catch and signal should have a variable gate which corresponds with the mean height of the fish above the sea bed, at different ranges, and for different sizes of fish. This point will be considered a little more fully below when the problems of catch correlations are analyzed.

The echo sounder was designed to give a continuous and accurate record of the depth of the sea. As noted in an earlier chapter, the depth of mid-water targets is nearly always over-estimated by a small percentage, dependent upon depth. So the range of a target and the signal expected from a target have to be examined statistically in the way set out in this section. The problem is not completely solved by using narrow beams because the mean angle or mean depth still has to be estimated. The variance of the distribution about the mean angle is reduced but so is the sampling volume. As a large number of signals is needed to obtain a good average, the advantage of reduced variance may be nullified by a loss of sampling power.

(d) *The volume sampled by the echo sounder*

It follows from the sonar equation that fish of different sizes are sampled at different maximum ranges, increasing with increasing size of fish. For any one fish size, it can be shown that the total volume is pear-shaped, increasing in diameter to a middle range and then decreasing to zero diameter at maximum range. For any range slice, between r_1 and r_2, there is a *range* shell volume defined by the ranges r_1 and r_2 and the angle θ_{min}.

For a square transducer, the range shell volume between r_1 and r_2 (meters) and out to θ_{min} in angle, is defined as:

$$\tfrac{2}{3}\pi \, (r_2{}^3 - r_1{}^3)(1 - \cos \theta_{min})$$

in cubic meters, and, for a rectangular transducer, by:

$$\tfrac{2}{3}\pi \, (r_2{}^3 - r_1{}^3)(1 - \cos \sqrt{\theta_\alpha \, \theta_\beta})$$

in cubic meters, where θ_α and θ_β are the angles subtended by the major ($r \sin \theta_{\alpha min}$) and minor ($r \sin \theta_{\beta min}$) axes of an ellipse at a range r. The range shell volume is that in which a signal from a fish can be detected. It can be split into sectors of angle θ. Each sector is multiplied by the appropriate directivity coefficient and the sum of all sectors, so weighted, is the effective volume sampled; it takes into account the chance of detection varying with directivity. The range shell volume is determined by the size of the fish and so when the average size is determined, the number of signals from fish and the average volume in which they lived is also estimated. Figure 60 shows the volume sampled in a single transmission for different sizes of fish in range.

There are two ways in which fish can be counted: (1) as number of traces per unit volume over a large number of transmissions, and (2) as number of signals per unit volume within one transmission. The first, trace counting, is used when fish are counted from a paper record and the second, signal counting, is used with automatic counters. In the first, the fish signal is recognized by the trace-to-trace correlation of successive pings. In the second, the fish signal is recognized merely by the signal-to-noise ratio in the single transmission;

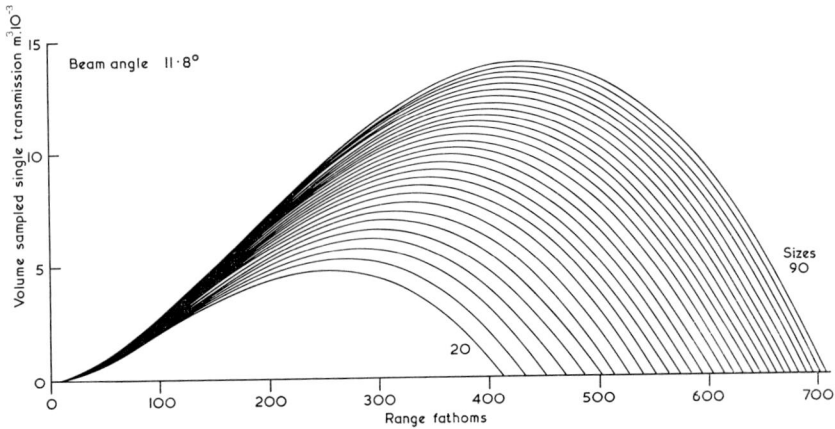

FIG. 60. The volume sampled in a single transmission in range for fish of different sizes, with a Kelvin Hughes Humber gear (Cushing, 1973).

hence in automatic counting, there is a need for frequent monitoring of noise and complete elimination of electrical interference from the ship. The volume used in the two methods differ. That in the signal counting method is the simple *range shell volume* given above. That in the trace counting method is more complex, being a projection of a single elliptical shell on to a vertical plane at right angles to the ship's course, which is extended along the ship's track for a given distance (see Harris, in the appendix to Cushing, 1968b).

The true volume, V_T is given by:

$$V_T = D \left[r_1{}^2 \, \theta_\alpha - \frac{r_2{}^2 \sin 2\theta_\alpha}{2} - \frac{R_x{}^2}{2} (2\psi - \sin 2\psi) \right] \tag{39}$$

where D is the distance steamed by the ship in meters; R_x is the radius of a circle passing through $r_2 \sin \theta_{\alpha min}$ on either side of the ship and $r_2 \sin \theta_{\beta min}$ on the acoustic axis; ψ is the angle subtended by the circle of radius R_x between the acoustic axis and $\theta_{\alpha min}$.

There are two errors in the estimation of V_T; due to discontinuity and to overlap. The discontinuity error arises because the edges of the beam do not always cover the true volume as the shells overlap. The overlap error arises because the range shell volumes in successive pings overlap in two dimensions, along the ship's track and vertically. The error varies with speed, pulse repetition frequency, and with depth. Figure 61 shows the volume sampled per nautical mile in range for different sizes of fish.

There are statistical differences between the two procedures. With signal counting the distribution is essentially the product distribution shown in Fig. 54. In principle it is easy to handle and can be used to extract size distributions from all the information received. With trace counting, the distributions are difficult to handle. The volume, V_T, may be considered as being halved by a projection of the acoustic axis along the ship's track and there is a distribution of signals along this projection. We cannot distinguish signals from the two halves nor should we assume that fish cross from one half to the other. But if it is assumed that the speed of fish is a small proportion of the ship's speed, half the frequency distribution of signals may be considered as being centered about the mean angle in half the volume, or all in the whole volume.

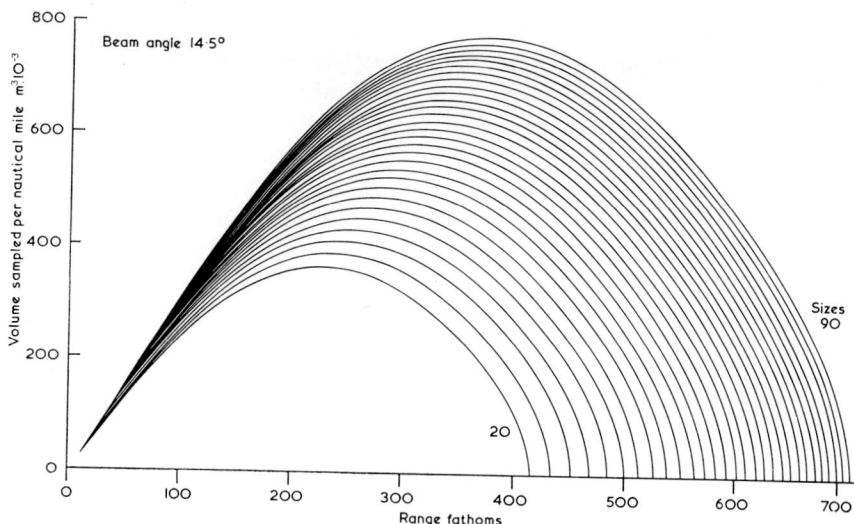

FIG. 61. The volume sampled per nautical mile in range for fish of different sizes, with a Kelvin Hughes Humber gear (Cushing, 1973).

One of the consequences of these calculations is that it becomes possible to calculate the expected number of pings from a given size of fish. At the mean angle the number of pings is given by the ratio V_T/V^1, where V_T is the true volume sampled by the echo sounder with the errors of discontinuity and overlap taken into account and V^1 in the volume sampled in the same distance steamed by a number of contiguous and exclusive range shell volumes. The ratio V_T/V^1 is calculable and could be used to compare with observed ping distributions. Not only is the number of pings per fish a measure of target strength, but it is possible that the differences in directivity of the swim-bladder, as suggested by Midttun (1966) might lead to discrimination between species.

In Anon. (1969) a method is given for estimating the area sampled at a given depth based on the average number of pings per trace. It was assumed that the beam was an elliptical section with the major axis athwartships and that the average chord track passed parallel to the minor axis at a point half way along the major one. On the ELAC echo sounder the ratio of minor to major axis was 2/31. Then the average chord track is given by $1852.V.\bar{n}/60.e$ where V is the ship's speed in knots, e is the pulse repetition frequency in m min^{-1}, and $\bar{n}$ the average number of pings per trace (which is observed) at a given depth. There are 1852 m per nautical mile, using the equation of the ellipse,

$$x^2/a^2 + y^2/b^2 = 1$$

where a and b are the major and minor axes respectively, where $2y$ is the average chord track, and $x = a/2$, a can be calculated.

Then the area bounded by the ship's track and the major axis of the ellipse is the area sampled by the echo sounder at a given depth. The method was evolved by Midttun to estimate the number of cod in the spawning layer in the Vestfjord in northern Norway; in fact he expressed the number of fish as per square cable or 1/100 of a square nautical mile. It is a less complex method than the more purely volumetric ones described above. Further it has the advantage that an estimation of size is unnecessary and that numbers alone are given, which may be all that is needed for some purposes. Monstad *et al.* (1969) have applied this method to the layer of spawning cod in the Vestfjord and the density of individual fish was expressed as *n* (between 1 and 200+) per nautical mile and as numbers per square mile. In the early part of the season the cod lay at 100–130 m and they rose to 80–100 m, roughly following the 5°C isotherm.

It is possible to calculate the number of pings per fish on an average track which passes equally about the mean angle. Figure 62 shows the number of transmissions per target at 1 knot and at 1 transmission per minute for different sizes of fish in range, with a Kelvin Hughes Humber gear. Little discrimination in size is possible at short ranges, but a fair discrimination is possible at greater ranges.

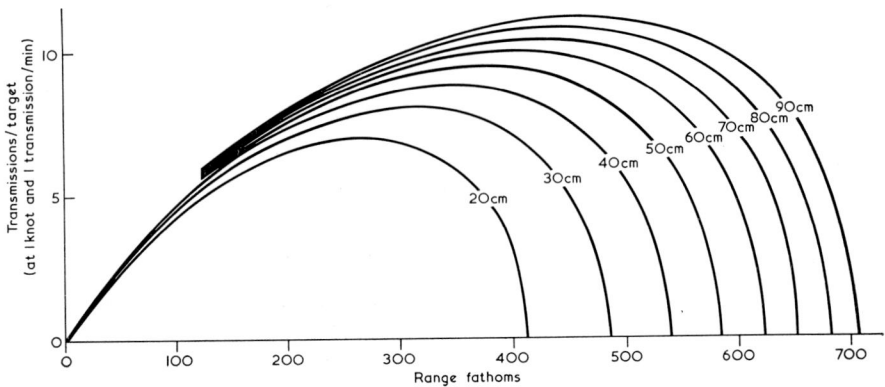

FIG. 62. The number of transmissions per target at 1 knot and at 1 transmission min^{-1} for different sizes of fish in range, with a Kelvin Hughes Humber gear (Cushing, 1973).

(e) *The estimation of density, exemplified by the hake survey off South Africa*

In February 1966, an echo survey for demersal fish was made between Cape Town and Walvis Bay aboard the Hydrographic Survey vessel, H.M.S. Hecla. The echo sounder used was a Kelvin Hughes Humber gear ($S = 128.9$ dB *ref.* 1 mcbar; beam angle, $9 \times 14°$ in echo) which has a bottom-locked, triggered pen recorder showing all signals received from a 4 fm scan above the seabed. On leaving Cape Town the noise level was measured in water deeper than 50 fm and the gain of the receiving amplifier was marked in a depth scale to detect a 20 cm fish at a conventional signal-to-noise ratio. The average amplitude was logged in each $\frac{1}{2}$ hr steamed, together with the gain setting and the depth. The number of single fish was counted in each $\frac{1}{2}$ hr steamed in a 1 fm scan off the bottom and in the whole 4 fm scan.

Figure 63 is derived from the average amplitudes observed. Most of the fish observed were hake. The mean height of fish off the bottom in the 0–1 fm scan is 0.75 fm and the height of a Granton trawl is probably 1.5 fm off the bottom at its highest point. So most of the fish observed in the 0–1 fm scan were vulnerable to capture by the trawl. Hake trawlers, working in three different areas, off Cape Town, south of the Orange River, and off Lüderitz in S.W. Africa caught 95% hake. So the signals in the 0–1 fm scan at least were nearly all from hake. In the 0–4 fm scan the evidence is less secure, although ships working off Lüderitz with high headline trawls (said to be 5 fm) were still catching nearly all hake, with some kingclip. The selection length for hake with the trawls used was just

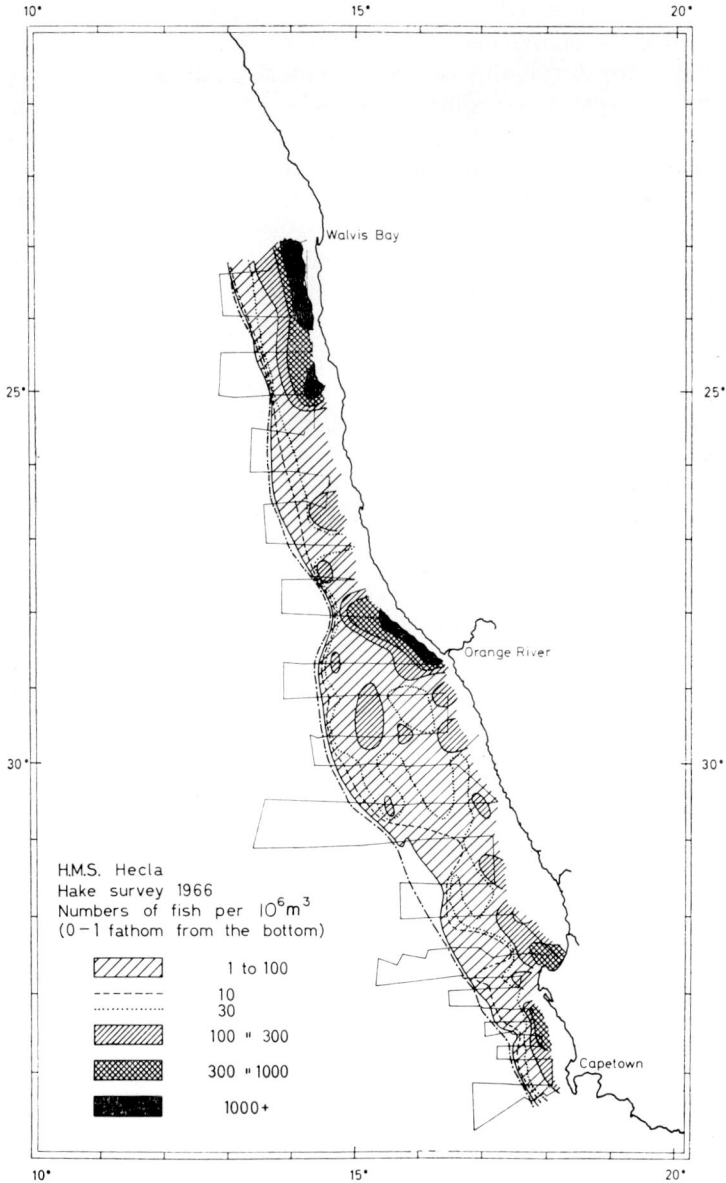

FIG. 63. Distributions of single fish in numbers per 10^6 m^3 in the 0–1 fm scan off the bottom in a survey for hake off South Africa (Cushing, 1968).

above 30 cm; it was shown above that the echo sounder was set to detect fish of 20 cm— so in a rough way, trawl and echo sounder were sampling the same populations.

A diurnal variation in hake catches is well known (Hickling, 1927) and on some occasions they have been caught in gill nets near the surface. The records were examined for such an effect and it was found that in the southern areas it did not exist, and that there was a slight variation with depth in the north. Perhaps the survey was executed at a season (summer or early autumn) when the diurnal depth variation was least.

Figure 63 shows the distribution in the 0–1 fm scan as density in numbers per $10^6 m^3$. The area of survey is spread across the continental shelf to a depth of 360 fm, which is the limiting depth of the machine which was in use. This was the first echo survey made in which the density in numbers of fish was shown and it is shown irrespective of the depth. Amongst other things it shows that the greatest densities of the smaller fish, of course, are found inshore. Effectively there are three nursery areas, north west of Cape Town, off the Orange River, and in the region south of Walvis Bay. The existence of three nursery areas pre-supposes three stocks and it would be very interesting if their existence were confirmed later. On two occasions it was possible to compare the acoustic density with the density of trawled fish; off Cape Town the estimated density from the echo sounder was 0.3 tons hr^{-1} as compared with an observed catch of 0.5 tons hr^{-1} and off Lüderitz the estimated density was 0.9–3.0 tons hr^{-1} as compared with an observed catch of 1.5–2.5 tons hr^{-1}. Hence, there is a little evidence that the density, as estimated by the echo sounder, is not far from that estimated by the trawl. If the trawl only took a small part of the stock one might have expected the acoustic density to have been higher—but it was not so.

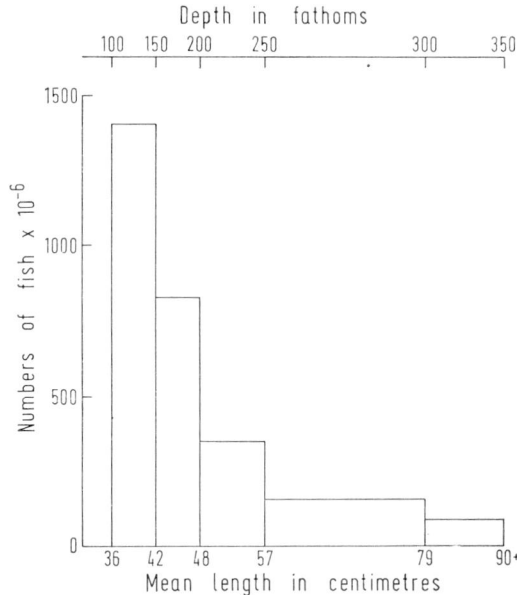

FIG. 64. A length distribution of fish sampled off South Africa using the distribution of density, in Fig. 59, and the dependence of size on depth, in Fig. 55 (Cushing, 1968).

An estimate of total numbers was made in the following way. A chart was drawn of numbers per 10^6 m^3 as at a mean length of 46 cm and expressed as numbers per 10^4/m^2 of surface. A bathymetric chart was overlaid and the numbers within a depth interval were summed. From the average line of mean signal on depth shown in Fig. 56, depth intervals were transformed into specified intervals of length. So, following the numbers per unit surface in depth contours, a distribution of the total stock was obtained. Figure 64 shows the length distribution so found; because the sampling volume appropriate to a fish of

46 cm was used, the distribution is biased. Fish smaller than 46 cm were counted in a volume appropriate to 46 cm fish and thus density was under-estimated; conversely the density of the larger fish was over-estimated. For this reason the length distribution in itself is of little value, but it should be clear that, with the right form of data processing, length distributions should be available even from the short time periods needed to derive an adequate estimate of the mean angle.

At first sight it would appear that most of the problems were solved in this one lucky survey. However, there are two which remain. The first is that because fish cannot be identified acoustically, the survey should strictly be limited to the zone of capture—in this case the fathom and a half above the seabed sampled by the bottom trawl. Then the purpose of the acoustic survey is to endow the trawl survey with an enormously increased sampling power and to provide an estimate of density independently of the trawl. This means that the trawl catch is used only to provide evidence of identification. The second problem is concerned with the mixture of species which is often found. The trawl catches are used to get length distributions of the different species and the length distribution of all fish is partitioned at each length group by species. So long as the proportions of different species remain the same between trawl hauls, the acoustic signals can be processed to yield length distributions of each species. If the proportions change between trawl hauls then the hauls must be set much closer together until the criterion of the constant proportion between hauls is fulfilled.

(f) *The estimation of density in fish shoals, as exemplified by the herring surveys in the Norwegian Sea*

In principle, it is possible to count and size individual fish acoustically. In practice, however, both single fish and multiple fish echoes may be received in the course of a survey. To be of general application it is necessary to count the single fish echoes and to estimate the number of fish in the multiple echoes. The latter is a difficult problem which has not yet been solved theoretically. Within one wavelength, the amplitude of the echo signal varies as the square root of the number of targets (or intensity varies with number of targets). It might be reasonable to extend this rule to one pulse length, because a pulse comprises a fixed small number of cycles. Further extension within a shoal requires the assumption that absorption of sound within the shoal is negligible. As fish shoals have been observed on rare occasions to cast shadows on the bottom (Cushing, 1963), indicating intense absorption, the assumption might be unreasonable except where the shoals are small ones. The effect of such absorption within the shoal will lead to under-estimation of numbers. Another effect is reverberation between fish. When fish shoals are near the bottom, signals are sometimes observed as a marked extension to the bottom signal, as shown in Chapter 2, implying that reverberation continues for a long time within the shoal. Its effect, if included in the voltage received from the shoal, would be to over-estimate numbers. It is concluded that the square root approximation may be valid for small or diffuse shoals (Richardson *et al.*, 1959), but may not be so for large or thick shoals.

In the U.S.S.R. this problem has been tackled by combining echo surveys with direct estimates of fish density (Truskanov and Scherbino, 1966)—with underwater cameras and with transducers lowered into the shoals at night. Charts have been prepared showing the distribution of fish density both horizontally and vertically. When fish are recorded as individuals, they are counted as such on the paper record, but shoals are treated in a fleet operation. The positions of the shoals of herring in this overwintering area are charted

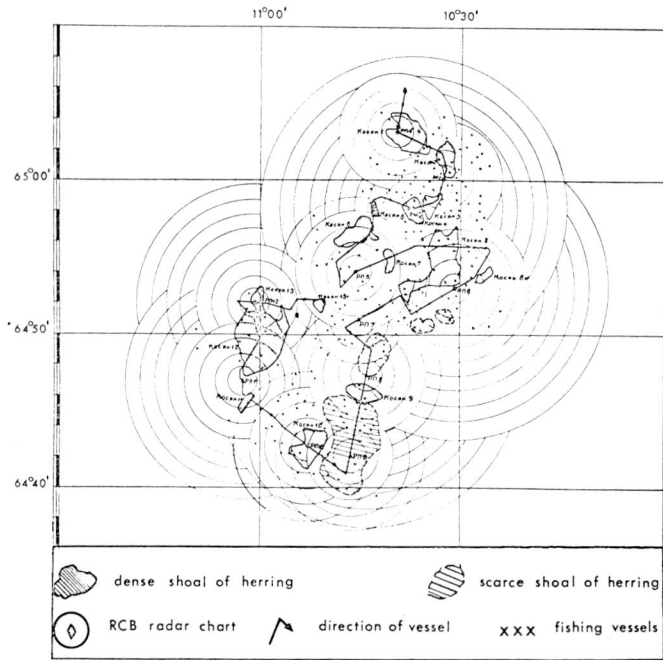

FIG. 65. Distribution of Norwegian herring in their overwintering area south east of Iceland in December (Truskanov and Scherbino, 1966).

with respect to the position of fishing vessels on a radar scan. So an instantaneous survey is made of an area several miles across by means of the fishing vessels and the research vessels. Figure 65 shows such a chart. Then the density is estimated in each shoal (a) using an underwater camera, (b) lowering a special high frequency transducer into the shoal, and

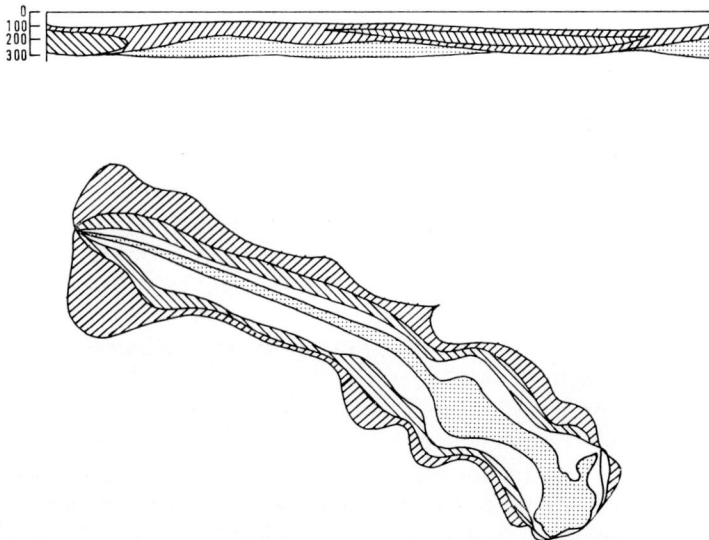

FIG. 66. The density of fish within a shoal of herring (Truskanov and Scherbino, 1966).

(c) from the average amplitude of the received voltage. Figure 66 shows the chart of density in numbers within a fish shoal and Fig. 67 gives the relationship between amplitude and density, where $V = 12.65 \sqrt{d}$, where V is voltage, and d is density in numbers. In view of the possible dangers in this procedure discussed earlier, the low variance about the line is

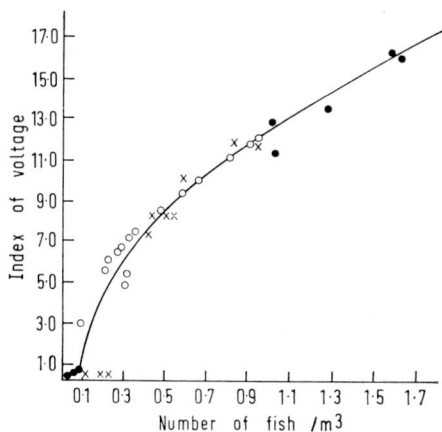

Fig. 67. The dependence of voltage on numbers in a shoal (Truskanov and Scherbino, 1966).

remarkable. However, the most important aspect of this work is in the estimate of numbers of the Norwegian herring stock. Truskanov and Scherbino (1966) give their results for a number of years:

	1958	1961	1962	1963	1964
Length of recording as % distance steamed	85	53	50	32.5	80
Thickness of shoals (m)	70	85	107	114	196
Area of concentration (km²)	260	142	267.2	20.5	239.5
Volume of concentration (km²)	18.5	15.5	21.3	20.3	23.7
Density (n 1000 m^{-3})	950	450	680	785	1,166
Stock in millions	17,575	6,975	14,485	16,092	27,640
Average wt (g)	344	359	196.5	202	246
Stock in weight (tons × 10³)	6,011	2,540	2,848	3,256	6,800

The figures correspond fairly well with the stock figures obtained by the traditional methods of fisheries research. One of the reasons for success was the discovery of an over-wintering area of the Norwegian herring in the East Icelandic current, south east of Iceland, in December. Surprisingly, the area is comparatively compact and, as it is on the main migration route and because the stock estimates are closely related to those based on the market measurements, it is likely that the whole stock of Norwegian herring was encompassed by these surveys.

Thus, provided that estimates of density are properly executed, the measures of stock in numbers can be made acoustically on pelagic fish as well as on demersal fish. In the future perhaps the two methods might draw towards each other; the statistical treatment would be of great value to the herring studies and true estimates of density to the studies of demersal fish. Indeed, it should be recalled that one of the first underwater photographs of fish in the sea was taken of cod in their spawning layer off the Lofoten Islands (Saetersdal, 1960).

(g) *Forms of data processing*

The paper record produced continuously by the echo sounder is a very adequate way of processing the very large quantities of data collected by the transducer. It is only since it became obvious that individual fish could be sized with an echo sounder that a real need for automatic data processing has arisen. The triggered pen recorder of the Kelvin Hughes Humber gear, which resolves single fish in time by the high speed recording of the scan close to the bottom, represents the ultimate presentation of the paper record.

The first requirement for any system of automatic data processing is that signals from individual fish be separated from those of shoals, because the amplitudes of the latter have to be transformed. At Lowestoft, a discriminator has been developed on the basis that the number of cycles received from a single fish is that prescribed by the transmitted pulse length, whereas that received from a shoal is greater. So the first stage in the processing system has two channels, a single fish channel and a shoal channel and each can be counted separately.

There are two forms of processing equipment in use with current echo sounders. The first is an integrator which sums all received voltages in a short time period on a paper record. It was first used by Dragesund and Olsen (1965) on the scattering layer of yearling fish in the Barents Sea, but it has since found additional applications. The second equipment is the cycle counter (Carpenter, 1967) which sums all signals received as total number of cycles received in a short time period; at the least estimate, the cycle counter gives a measure of the area on the paper record covered by a fish trace. In a more sophisticated way, with the single fish discriminator, the number of cycles received within a range gate for a short time period is an index of the number of single fish encountered; the total number of cycles is divided by the number of cycles per pulse and the number of pings per fish (given by the ratio V_T/V^1), appropriately for a particular size of fish; it will be recalled V_T is the true volume sampled in the same time period by a number of contiguous range shell volumes. The figure printed out then represents the number of fish of a given average size (which determines the volumes sampled) in the range gate for a time period.

A useful combination is to use discriminator, integrator, and cycle counter together. Within a range gate, for a short time period, the discriminator separates single fish signals from shoal signals. The single fish channel is counted by the cycle counter giving a print out of the number of individuals which can be expressed as numbers per unit volume. The shoals channel voltage is summed by the integrator. The total voltage divided by V_T/V^1 and transformed by the square root (so long as the shoals are not too dense) should give the number of fish in the shoals. The integrator finds a special use in the study of scattering layers where indeed it was first used by Dragesund and Olsen (1965). It will be recalled that the volume reverberation coefficient is calculated in a sonar equation using $20 \log r$ rather than $40 \log r$ in the one way propagation loss. Hence, if the voltage is squared before integration, the integrated voltage will correspond to the number of targets, expressed as a volume reverberation coefficient. With a time-varied gain, the figures can be used directly (with $20 \log r$ in the transmission loss) to compare the living material in scattering layers at different depths and at different times.

Thorne and Woodey (1970) assessed the stock of juvenile sockeye salmon in Lake

Washington, near Seattle in Washington State, U.S.A. A Simrad EH 2E (at 38 kHz) was used with an integrator and a time-varied gain with a 20 log r law (i.e., volume reverberation). An initial calibration was made with Isaacs-Kidd trawl hauls of small, medium, and large fish from different depths; a multiple regression was calculated of integration rate on the numbers of the three size groups of fish. A very full series of transects across the lake at night was carried out at a number of depths and catches were made at the same time; a proportional regression was calculated of numbers on integration rate and the variance was weighted by the square of the integration rate. The numbers of fish were 8.92×10^6 and the error of this estimate, expressed as two standard deviations, was 1.43×10^6. The coefficient of variation is low, 16%; to obtain the same result by conventional methods would be very expensive in time and money. It has been suspected for a long time that the methods of acoustic survey should be cheaper and more accurate than survey by exploratory fishing, but this is the first demonstration.

A simpler form of counting equipment has been developed by Hearn (1970) for use in the commercial fishery; it counts pings as from single fish in scans of 2, 4, 8 or 24 ft off the bottom with a Kelvin Hughes Humber gear. Fish larger than 25 cm are counted and a correction for range is made with a smoothing constant; the catch is estimated to within 50% for 60% of the time. Shibata et al. (1970) have described an echo counting equipment with a time-varied gain which counts the number of pings per fish as at a constant ship's speed. An optical integrator measures the areas of the traces of fish shoals as recorded on paper record. Suomala (1970) developed a computer program which expresses the distribution of signals as a plan picture along either side of the ship's tracks. These methods are useful developments of echo counting and express the need felt by fishermen and fisheries biologists for the automatic counting of fish signals.

A quite different procedure has been devised by Dowd (1969) for a specific purpose. To examine the fish in the path of a trawl for demersal fish, a transducer is "flown" at a fixed and constant height above the seabed, at such a height that the scan corresponds roughly to the area swept by the trawl. The great virtue of this system is that the number of individual fish in the path of the trawl is counted simply, with none of the statistical complexities raised in an earlier section. The fish are recorded as pulses which are stored for the period of the trawl haul. Some preliminary results show a clear correlation between catch and signal, which is needed for a study of catchability in progress at the New Bedford Institute of Oceanography.

The most sophisticated equipment at the present time is that developed at the Marine Laboratory, Aberdeen (Craig and Forbes, 1970). A transducer working at 400 kHz is designed to split herring shoals into arrays of individual fish in the surface waters of the sea off Scotland. The four hundred channels of a pulse height analyzer were divided into two groups, a series of range gates in depth, and a series of channels in amplitude. So a frequency distribution of signals is generated in a short time period at each range gate. The mean of this distribution is the signal received from the mean angle and the system is well designed to take advantage of the statistical characteristics of the echo sounder.

The quantity of data produced by such equipment obviously needs computer handling. There are three requirements to be satisfied. The first is that a time-varied gain be used; preferably the control of the gain should be set at the maximum range of the smallest fish desired in the sampling system, which of course depends upon the noise level measured and the signal-to-noise ratio chosen. The second requirement is that all the data received are fully used. The mean of the frequency distribution of signals is that at the mean angle for the average size of single fish in the range gate; so the product distribution is calculable at all angles in the beam for fish of that size. This distribution must be calculated for each short time period as the average signal is computed. Then deviations from the distribution are functions of the true length distribution of the fish, so a transformation is needed to convert deviations into true amplitude and true length. The third requirement is a ship borne data logger which processes all the received signals and expresses them as length distributions for each short time period in each range gate.

The recording echo sounder was an early form of analog data processor. The various forms of equipment which have been described thus have limited purposes. The cycle counter, integrator, and pulse counter are in use to count fish as single fish or as shoals. But to be of real use in fisheries research as an aid to population analysis and in exploration, the equipment needs to be fully developed.

6. The use of low frequency equipment

A recent development shows that fish can be detected at long ranges (tens of nautical miles) and that their average densities can be estimated across such ranges. Weston *et al.* (1969) studied the propagation of sound at low frequencies (1–3 kHz). One of their more striking results (Weston and Revie, 1971) is shown in Fig. 94, in a later chapter. Their estimate of fish density, however, is based on the observation that attenuation varies diurnally; indeed it dropped 30 dB 40 min after sunset. In Fig. 94 the fish traces disappear at night. They suggest that attenuation $d = 10 \ln N\sigma_e$, where N is number of fish per unit volume and σ_e is the extinction cross section (in m²). At such frequencies the swim bladders of fish resonate, then $\sigma_e = \lambda^2 Q/\pi Q_r$, where Q is the overall Q factor and Q_r that due to radiation damping alone. Ching and Weston (1971) estimated densities at three frequencies:

kHz	0.7	1.44	3.55
L (cm)	24	12	5
a(dB km^{-2})	1.62	1.62	1.08
σ_e (m²)	0.117	0.028	0.0047
N m^{-3}	3.18×10^{-3}	1.33×10^{-2}	5.30×10^{-2}

It is assumed that fish are not concentrated at the surface or on the bottom and that the source of attenuation is resonance of the swim bladders of fish which shoal during the day. It was suggested that the fish of 24 cm were pilchards. If the swim bladders are resonating, the use of different frequencies would obtain discrimination in size. In principle then there is a method for sizing and counting fish across distances of tens of km.

7. The correlation of catch and signal

An assumption in fisheries research is that the capturing gear catches a fixed proportion of the population; ideally a trawl catches all the fish in its path as if they were marbles. But fish escape and the quantity caught in the area swept may only be a fraction of the quantity living there. The fraction is unknown and it may vary with night and day, with season, or with depth. All stock assessment is based on the catch of the fishing gear because the average catch represents a proper estimate of stock density and indeed the catchability coefficient in $Z = M + qf$ (where Z is the instantaneous coefficient of total mortality, M

is the instantaneous coefficient of natural mortality, f is fishing intensity, and q is the catchability coefficient) represents the fraction caught of the quantity living in the area swept. The requirement is to assess the fraction independently of the market measurement system by which stock parameters are estimated at the present time.

The first indication that catches of fish might be correlated with signals from fish, and hence might be predictable, was found in the work of the late Skipper Balls of Great Yarmouth. He showed that shots of herring in his drifter could be predicted more than thirty years ago. By the early fifties, most herring fishermen in the north east Atlantic and north east Pacific were using echo sounders to find fish. During the late fifties and early sixties, the purse-seine skippers from Norway and Iceland developed a system of aimed fishing with sonar, which implies some sort of prediction of quantity. The same form of prediction can be made by the best trawler skippers, as shown by the recently published work by Drever and Ellis (1969).

The explicit quantification of the skills of fishermen came very slowly. One of the first correlations was made by Richardson et al. (1959) who showed a relationship between catch and signal for about forty trawl hauls on cod in the Barents Sea; a rough correction was made for depth by correcting the signals for the transmission loss. Mitson and Wood (1962) showed a very clear correlation between catch and signal at the same depth using automatic counting equipment. For the drift net fisheries for herring off East Anglia, Burd (1964) (Fig. 73) published a correlation between the average catches of drifters and the average quantity of traces observed on echo survey over the area of catching. All these correlations were successful because a wide range of catches was available. When the same technique has been applied by fishermen it has failed. It failed in the East Anglian herring fishery because the variance of single catches is very high; it failed in the distant water cod fisheries because the differences in catches were not great enough. Yet a successful correlation has already been noted, that published by Dowd (1969). The reason for this success is that differences in volume sampled by the echo sounder were eliminated by flying the transducer at a fixed height above the bottom. When it is recalled that there is a specific sampling volume for each size of fish at each depth, the source of high variance must lie in improper estimates of sampling volume.

In an earlier section the methods of estimating sampling volume were reviewed; essentially, the signals from a range gate for a short time period are averaged and from this average, the size of fish is estimated, and a sampling volume allocated. In a drift net fishery or a long line fishery which is near the surface and not depth-dependent, the density of fish so derived is adequate. In a trawl fishery there is a special problem in two parts: (a) the ratio of volume sampled by the echo sounder to that sampled by the trawl varies with depth, being a function of mean angle between the transducer and the maximum range specified for the average size of fish, and (b) the trawl samples a different volume, a projection of the mouth section along the ship's track, whereas the echo sounder samples a sector between two spherical waves as specified by the range gate. If fish are randomly distributed, only the first part of the problem matters, the ratio of volumes sampled at different depths. However, demersal fish are often distributed in layers and then the problem is to adjust the range gate to sample such layers most effectively.

The same method can be used to solve the first problem. It might be thought that a one fathom scan would approximate to the height of the trawl headline sufficiently. But eqn. (38) gives the mean height of fish above the bottom if they are distributed about the mean angle in the beam. As shown in Chapter 3, the mean height is a function of mean angle, itself a function of size of fish. So the one fathom scan on the recorder represents a height above the seabed which varies with depth and with size of fish. Figure 68 shows the trend in the one fathom scan as height above the bottom depth and for size of fish. Thus a fixed scan on the record or on the counting equipment over-estimates the volume which matches the trawl volume at middle depths and under-estimates it at extreme depths, for size of fish. The problem is best reformulated. Given the size of fish and hence the mean angle, the mean height above the bottom in a range gate is calculable. So the range gate needed to approximate the mean height to the height of headline is also calculable.

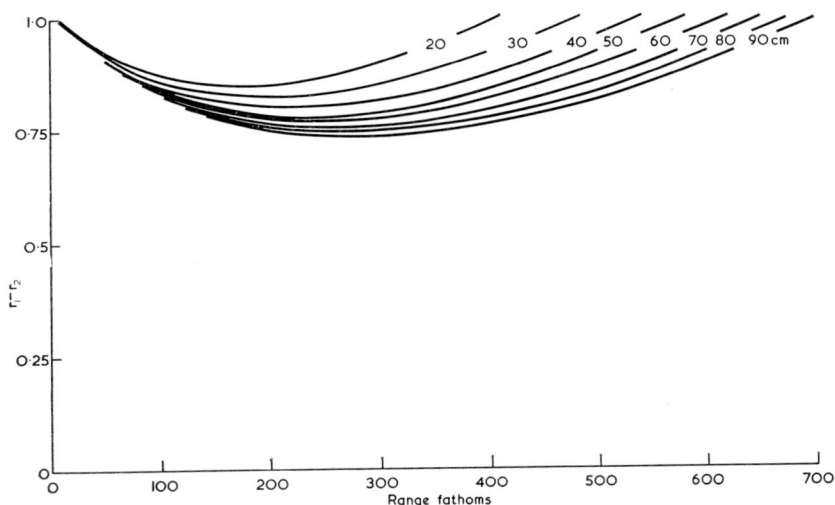

FIG. 68. The scans in fathoms on the display (oscilloscope or paper record) for different sizes of fish in range for a Kelvin Hughes Humber gear (Cushing, 1973); $h = 0.5$ fm.

Captain Drever and Mr. Ellis (1969) have pointed out that fish are often hard down on the bottom, the signals only being recorded in about a quarter fathom above the seabed. This is because the spherical wave of the sound pulse touches the bottom at one point only and the bottom signal returned from that point masks the presence of fish at any angle to the axis if they are very close to the bottom. With a flat bottom, a cod on the axis should be detected; but if the bottom is less than flat, fish may disappear in the roughness and not be detected at all. Hence, there is a chance of catching fish with no signal recorded on the paper and, with those recorders without a white line or with a coarse scale, the chance is fairly high and fishermen have at times lost confidence in such equipment. With more advanced machines the chance of catching fish without detecting them at all is much reduced. If, however, the fish are detected in the lowest quarter fathom of the paper record, then it is obvious that the echo sounder volume is represented by a very small spherical cap and that this can be estimated properly. Consequently the range gate must then be set to count fish in only that narrow volume. Thus there is a two stage procedure for reducing the volume variance in a correlation of catch and signal. With the usual procedure, the size of fish is determined and, from the mean height above the bottom at the chosen depth, the right range gate is estimated. If the fish are randomly distributed within that gate on the paper record, the count proceeds as from that gate; if, however, they are layered the range gate is adjusted to the height of the layer and the volume recalculated from the new estimate of $(r_1 - r_2)$. If this procedure is followed then the sources of volume variance have been eliminated from the correlation of catch on signal.

There are three reasons for investigating the dependence of catch on signal: (a) to obtain estimates of stock density independently of the average catch of fishing vessels, (b) to be able to predict the catches of fishing vessels, and (c) to investigate changes in the catchability coefficient in detail.

8. *The use of the estimate of absolute abundance in fisheries research*

The study of populations in fisheries research is primarily based upon the average catches of fishing vessels which, in theory, provide good estimates of stock density. The statistics of catches and time spent fishing (or fishing effort) collected in the ports are augmented by large quantities of market measurements; on markets in England and Wales, each year, about 0.75 M fish are measured in length with a comparatively few age determinations, and with them the length measurements and the statistics of catch and effort are converted to age distributions in stock density. Such estimates can be used on a year-to-year basis to provide measures of growth and mortality. With tagging experiments at sea of various forms, information of this kind is used to construct population models by which the stocks are controlled internationally.

But this simple view of the population dynamics of stocks of fish is a little misleading for two main reasons. The first is that the fleet may not fully exploit the range of the stock; if only part of the stock is available in area, that part is said to be accessible and if all the stock is accessible but not caught by the gear (for example above the headline of the trawl) it is said to be vulnerable to the gear. Fish may be vulnerable to some gears but not to others, accessible at one season, but inaccessible at others. Vulnerability and accessibility are terms describing measurable modifications of the catchability coefficient. Hence the catchability coefficient must be well described if the biases in its estimate are to be analyzed and measured. The second reason why the simple view is misleading is that the catchability coefficient is often difficult to measure properly. Worse, it takes a long time to do so, perhaps many years of observations.

Until the last decade or so, fisheries developed rather slowly and the increase of mortality with the increase of effort was one of the two main sources of information. The second source was from tagging experiments by which the effect of fishing could be measured. Both were used to estimate fishing mortality and hence the catchability coefficient, but one decade was a short time in which to collect enough information. But in recent years, fishing vessels have developed the range and power to move from stock to stock in the course of a single voyage. Russian trawlers fishing for hake off South Africa have steamed across the ocean to fish for hake off South America. British trawlers working off Iceland may move to Labrador or west Greenland or even to Newfoundland to catch cod within a single trip of two months. The Scandinavian purse-seiners have moved from Iceland to Jan Mayen, from Jan Mayen to the Barents Sea, from the Barents Sea to the North Sea, changing their catches from herring to capelin, capelin to mackerel. Purse-seiners working off Peru built up the catches of anchoveta from 0.1 M to 10 M tons in much less than a decade. Under such rapid development, the job of the fisheries biologist is made very much more difficult.

However hard the problems facing him, they must be solved. One way is to use acoustic methods to estimate the fish populations. They are not yet fully developed but there are a number of advantages in their use: (a) in a survey like the hake survey off South Africa, fishing mortality is given by the ratio of catch to stock, provided, of course that the echo sounder and capturing gear sample the same sizes of fish; (b) in detailed experiments, the catchability coefficient can be estimated directly, as Dowd has already shown in principle; (c) in exploration, an estimate of stock can be made before fishing starts, which is a very great advantage; and (d) with repeated surveys of an exploited stock many independent estimates of fishing mortality could be made in the course of a year, which would be analogous to a series of tagging experiments.

The general advantage of acoustic methods is that estimates of stock density are made independently of the fleet activity. Further, the work of exploration is directly linked to the work of the population dynamicists.

CHAPTER 5

THE STUDY OF FISH BEHAVIOR
ACOUSTICALLY

1. *Introduction*

THE study of fish behavior is of considerable importance to fisheries biologists. For example, the mechanisms by which fish migrate hundreds of miles in a year remain undiscovered. But the migration circuits from spawning ground to nursery ground and from feeding ground back to spawning ground are often well known. The first stages of population analysis require that the grounds on which fish spawn, grow up, and feed, be located and the migratory routes established. It is only through understanding of the migration patterns that the unity of the population can be comprehended. It is established, on a genetic basis, from the distributions of alleles in blood groups sampled on the spawning grounds, but the fisheries on all grounds have to be attributed to stocks. Hence, the population biologists need descriptions of the migratory circuits of different stocks as well as the power to distinguish them.

Direct observation, which has been established with echo sounders, has shown that the spawning shoals are larger than the feeding ones. Observations of this kind provide the first step towards elucidating the migration circuit. Indeed, North Sea herring trawlers, working in the late thirties without echo sounders, and after World War II with them, discovered the precise positions of many of the herring spawning grounds in the North Sea. By now, as shown in Chapter 1, fisheries biologists from Norway and Russia have charted the whole migration circuit of the Norwegian herring (Devold, 1951; 1952; 1963; 1966; Truskanov and Scherbino, 1964); others from Norway and Britain have charted a fair part of the migration circuit of the Arcto-Norwegian cod (Bostrøm, 1955; Midttun and Saetersdal, 1957; Saetersdal and Hylen, 1959; Hylen *et al.*, 1961; Richardson *et al.*, 1959). Thus, on a considerable scale, acoustic methods have been used to unravel the circuits of migration.

Another field in which acoustic methods could be used effectively is in the study of the vulnerability of fish to various forms of gear. Fishermen believe that they catch all the fish available to them, but some fisheries biologists believe that the gear may catch only a fraction of the fish available. Without rehearsing the evidence for the vulnerability of a fish stock to gear, it can be illustrated; for example, a hook with a fish on it catches no more fish, and a trawl catches no fish which swim just above its headline.

From the procedures described in Chapter 4, the sizes and densities of fish can be estimated directly above and below the height of the headline above the seabed. Thus the fraction available can be estimated as the ratio of the quantity below the headline to that above it. There are difficulties in this work, one, for example, being the uncertainty of the precise identification of fishes above the headline without capture. However, despite the difficulties it might become possible to make some progress towards the direct estimation of vulnerability.

The study of fish behavior has its own discipline and requirements which are independent of and different from the needs of the population dynamicists. The echo sounder is exploited by the behavior biologists to find how fish move and how they respond to varied stimuli. It is one of the tools used amongst others and most students of behavior would not be satisfied with the acoustic information alone; the echo sounder might show unequivocally that the fish were above the headline of the trawl, but, even if adequately identified, the echo sounder cannot be used to show what the fish were actually doing, how they were behaving in detail. Most students of behavior would want to see the fish. However, this chapter is not a study of fish behavior but merely lists the limited information available from an echo sounder, the direct observation of diurnal changes, shoal sizes, fish speeds, fish densities, and differences due to availability changes.

2. *Diurnal rhythms*

Because fish can be observed continuously with an echo sounder in the same area, the change in behavior from day to night was one of the first observations to be made. It had been known for a long time that there was a diurnal variation in catches of trawled herring (Lucas, 1935), of trawled hake (*Merluccius merluccius* Linnaeus) (Hickling, 1927), and of drift net caught herring. But because the fish were not caught by other gears, at the time of capture, the cause of the diurnal variation could not be established. Observations with the echo sounder showed how fish rose towards the surface at night and descended to a day depth (or to the seabed) at dawn; in other words, the cause of the diurnal variation in catches was a vertical migration which had been well described for other forms of marine animals which were adequately captured at all depths (Cushing, 1951). It would be still difficult to study the vertical migration of fish diurnally with a combination of bottom trawls and mid-water trawls, because such gears cannot be closed.

One of the first descriptions of a vertical migration of fish with an echo sounder was that made by Richardson (1952) on the sprats (*Sprattus sprattus* Linnaeus) of the Thames estuary in southern England. The layer of sprats was extensive and the ship remained in contact with it all night. Figure 69 shows that it rose to the surface at dusk and that it descended to a day depth after dawn. Richardson also compared the diurnal behavior of herring off Yorkshire, off East Anglia and off Calais on the French coast of the Straits of Dover. He showed that the depth of the herring traces was correlated with the depth of the Secchi disc in the pre-spawning shoals off East Anglia and in the spawning shoals off Calais; indeed the latter sunk a little from the surface waters in full darkness and rose at dawn before descending to their day depths. The shoals off the Yorkshire coast were spawning and their ascent at night appeared to be restricted by the depth of the thermocline, although a few shoals were shown to lie right across it. Valdez and Cushing (1966) have examined the diurnal variation in echo traces in autumn during the late fifties, in the Southern Bight of the North Sea; from stock estimates, they concluded that 98% of the echo traces were composed of herring. The average vertical migration was slight, from a depth of 9 m in the day time to 5 m at night; as the drift nets extended below this day depth, the fish were vulnerable to capture in day time. Although catches were made during the day, the great majority were made at night, which suggests that the herring may have seen the nets during daylight fishing.

In some cruises the amplitude of vertical migration was greater, particularly in waters around the Dogger Bank, but in two cruises, in late October and early November, there

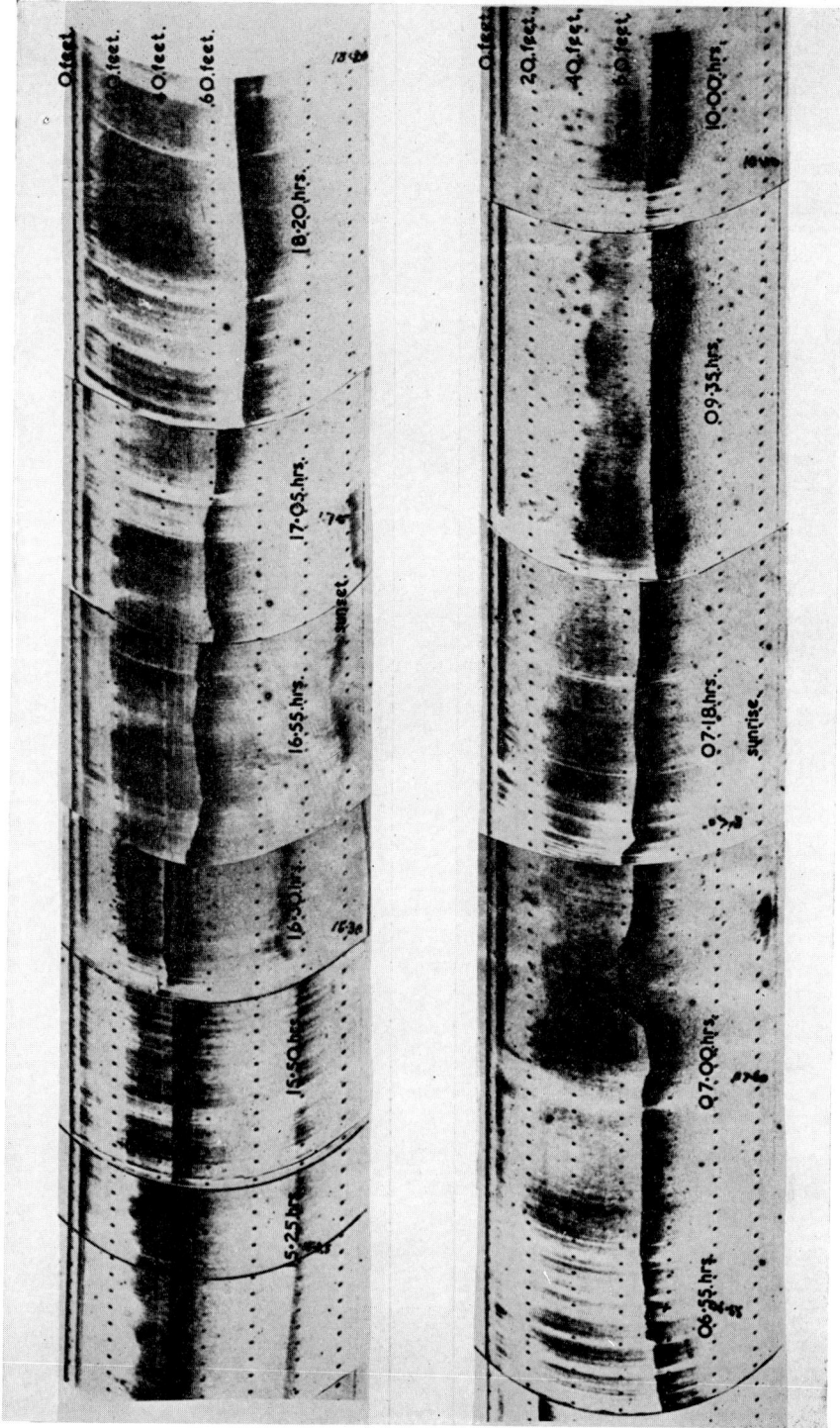

FIG. 69. Vertical migration of a layer of sprats in the Thames estuary (Richardson, 1952).

appeared to be no vertical migration at all. Figure 70 shows a vertical migration of herring in the Farn Deep off N.E. England (Postuma, 1957); the day plumes in the lower part of the water column rise towards the surface at night where they disperse into a layer just below the thermocline. This spreading pattern of behavior was named the "wineglass" pattern by the late Skipper Balls (Balls, 1951).

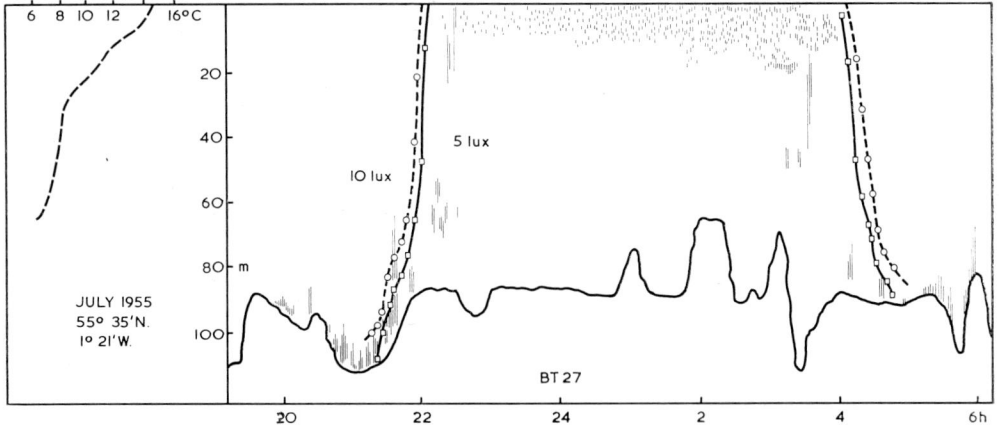

FIG. 70. Vertical migration of herring in the Farn Deep, in the North Sea, in summer time (Postuma, 1957); the plume traces of herring in the lower water column rise towards the surface and disperse into a layer below the thermocline.

The same type of migratory pattern was demonstrated for the sardine-like fish. Off Morocco, the sardines lived at a day depth of 25 m and at 1800 hr they rose to 5 m in the course of a few minutes (Furnestin, 1953). In the Adriatic, *Sardina pilchardus* climbed to the surface at night, but were restrained by a thermocline of 4.5–6.0° in 2–5 m (Zupanovic, 1967). Zei (1967) has shown that a layer of *Sardinella aurita* rises at night off Ghana in West Africa as far as the thermocline, but that the fish swim through the weak thermoclines found during the upwelling period. Thus the behavior of sardines in their vertical migrations resembles that of sprats or herring; the sprats and sardinellas tended to remain in layers in the daytime, whereas the herring which retreat to deeper water aggregated into discrete shoals. Both herring and sardines appear to swim through weak thermoclines and to be restrained by intense ones.

In Passamaquoddy Bay in New Brunswick, Canada, Brawn (1960) with records from many years, has shown that from May to September, the yearling herring remained at 10–14 m during both night and day. In the first quarter of the year, however, a marked vertical migration occurred between 50 m in the daytime and 10 m at night. In September, an inverted migration occurred when the fish lived at about 7 m in the day time and at 20 m at night. There appears to be no reason yet for this aberrant behavior. Another aberration was noticed by Harden Jones (1962) who found that some herring shoals on their spawning grounds at Sandettié in the Straits of Dover did not rise and disperse at night but remained in contact with the bottom, against a fairly strong tide. It was not shown decisively that the fish were spawning, but this appears to be a likely explanation. Because herring spawn on a gravel seabed of limited extent, it would be reasonable to suggest that they remained in contact with it, for the period of ten days or so during which spawning takes place (Ancellin and Nédelèc, 1959).

A number of observations have been made on gadoid or cod-like species. Schmidt (1957) showed that large cod shoals off Iceland lived on the bottom at dusk when they split into small shoals and then dispersed into a continuous layer 10–20 m thick; at dawn, this layer broke up into the compact shoals as fish aggregated on the bottom. In the Barents Sea, in September, Ellis (1956) found that cod concentrated in shoals in the daytime and dispersed at night, but at the same depth, so no migration took place. Sundnes (1964) has shown that a purse-seine does not disturb the spawning cod in the Vestfjord as recorded by an echo sounder; in other words, they did not move up or down and neither did they disperse. The diurnal movement is one of a layer which becomes denser and sinks a little in day time, and rises and disperses to a layer of individuals at night. Off the St. Lawrence and off Nova Scotia, cod and haddock were found on the bottom by day and in the mid-water by night (Beamish, 1966), so they can migrate vertically. Konstantinov (1964) demonstrated the same effect in the Barents Sea and in addition showed that the greater the diurnal difference in light, the greater the amplitude of their vertical migrations. At 76° N, he found cod descending at 60 m hr^{-1} at 0800 hr before any gleam of light appeared in the southern sky. Either the fish sank from the surface as they swam more slowly in the darkness or they were responding to very low light levels indeed, which seems unlikely in view of Konstantinov's observation.

Vestnes et al. (1965) studied the Chilean hake and showed that they rose and dispersed in irregular and partly broken layers of fish and then dispersed throughout the mid-water as individuals. A very clear description of the behavior of the Pacific hake is given in Alverson (1967). In the daytime, the shoals lay just off the bottom, sometimes in apparent contact with it, but at other times the lower part was as much as 2–20 m off. Between 1800 and 2100 hr the shoals rose and by 2300 hr they lost their identity and the fish had scattered. By dawn, the fish descended and began to reform near the bottom, but not necessarily in the same region. But the spawning shoals of hake off Port Susan, in Puget Sound (Washington State, U.S.A.) or off southern California, did not migrate vertically very much. Those examined off South Africa in February 1966 (Cushing, 1968) also did not appear to migrate very much.

The behavior of the gadoids resembles that of the clupeids in that a vertical migration appears to be the rule and that spawning fish are perhaps less responsive to the changes in light intensity which appear to mediate the behavior pattern. However, the gadoids live in deeper water, being bigger, and therefore must respond to lower levels of light intensity. So the range of migration is greater and the changes in shoal pattern appear to be more diverse if only because they take place rather more slowly in deep water than in shallow water. Perhaps more precise studies on light sensitivity and on migration with respect to light penetration should be undertaken to elucidate the difference within seasons and between species.

Similar studies have been made in fresh water, but less intensively. Haram (1965) has described the vertical migration of two layers of individual fish in Lake Bala, in North Wales. Johnson (1961) showed that the sockeye salmon *Onchorhynchus nerka* (Walbaum) remained between 3 m and the surface in the daytime and that it rose to the surface at dusk. But the most thorough-going investigations were made by Hasler and Villemonte (1953) in Lake Mendota, Wisconsin, U.S.A., using divers and echo sounders. Before sunset, perch *Perca flavescens* of 15–28 cm in length aggregated in tight shoals with 20–25 cm between individuals. So long as there was light for the divers to see with, there were shoals off the bottom. At dusk, the fish settled to the bottom where they dispersed with their pectoral fins resting on the bottom. At daybreak, the fish were seen to rise from the lake bed and to gather in shoals. The investigation was an echo-sounder study supported by divers, who identified the fish and examined the details of their behavior, when their positions had been revealed by the echo sounders. The point of identification cannot be pressed to the same proper degree for the other investigations described, but in each case a rough identification by catches was possible.

One of the interesting generalizations from many of the results is that fish tend to disperse at night. Theoretical work on shoaling shows that the schooling advantage with respect to predators is nullified at short ranges and so dispersal in darkness, or in turbid waters, is expected (Brock and Riffenburgh, 1960). Olson (1964) applied the theory of submarine warfare to the same problem and showed that the schooling advantage is less if the probability of detection were low. To increase the probability of detection, the area swept by the detectors must be increased and so one would expect that predators should school rather loosely: hence the observation that the daytime shoals of cod are smaller and more variable than those of herring is to be expected.

A fairly full description of the vertical migration of a few fish (and the associated diurnal rhythms) has been made with the echo sounder in regions where the question of identification is answered by persistent commercial catches. The area of abundance is usually extensive enough for the dominant species to be obvious for a period of time during which the echo sounder is continuously recording the vertical migration.

3. The effect of artificial light on shoals of fish

In the early days of echo-sounding research, attempts were made to modify the behavior of fish by switching on lights at night from a ship at sea. Hodgson and Richardson (1949) and Richardson (1952) showed that the depth of a fish layer could be modified by switching on an Aldis lamp as it was pointed down into the sea. A neat demonstration is given in Fig. 71, published by Verheijen (1959). Krefft and Schubert (1950) showed that layers of small whiting (*Merlangius merlangus* Linnaeus) and of *Belone belone* Linnaeus were driven downwards by light at night. Dragesund (1957) studied the extended response of herring: first, they descended and packed, after which they scattered and then the shoal would reform. Hence the response was a complex one and not a simple flight reaction. Indeed Kuroki (1969) has shown that fish (*Engraulis japonica* Houttuyn, *Trachurus trachurus*, *Scomber japonicus*) distributions were affected by the color of light; the orange light was at the near edge of the shoal and the far edge was in the blue light. Blaxter and Parrish (1958) have demonstrated that fish aggregated beneath lights at night; whiting gathered at 0.1 lux, whereas sprat and herring aggregated at 1,000–10,000 lux. This evidence recalls the difference between clupeids and gadoids in the amplitudes of their vertical migrations, the first shallow and the second deep.

There are two purposes in this work, first, to try to aggregate fish and secondly, to investigate the response of fish in darkness to a light beam. The two are linked, in that such behavior is unnatural and in that any rigorous study of aggregation in artificial light depends upon an analysis of behavior.

Nikoronov (1959) has developed a fishery for young clupeids in the Caspian Sea, in shallow water at night, which depends upon their aggregation under strong lights; the engineering structure is such that the fish are forced by the arrangement of lights into the mouth of a pump which sucks them up. The behavioral mechanism of the Mediterranean

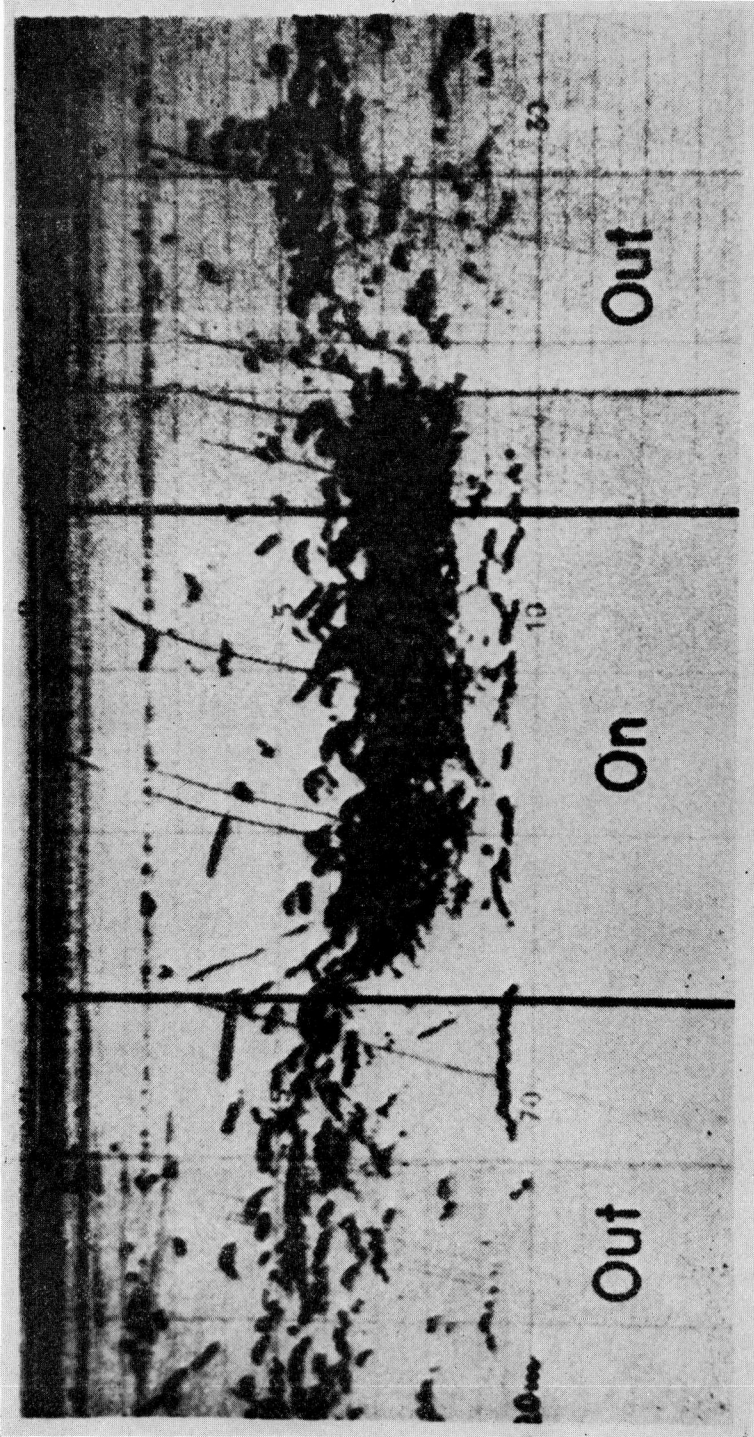

Fig. 71. The effect of artificial light on a shoal of sardines (Verheijen, 1959).

light fishery at night for clupeids has been investigated thoroughly by Verheijen (1958). He showed that, in very clear water, the differences in light intensity between the strong point source of light and the surrounding field are very great because a very small proportion of the light is scattered. Experimental work demonstrated that the fish are forced to move in the direction of the light source if the intensity differences across the visual field are large (three or four orders of magnitude); no forced movement occurs when the intensity differences are low, as would be expected when a large part of the light is scattered. So the light fishery in the Mediterranean depends upon the lack of scattering of light in the clear water. It is not known whether the Caspian Sea fishery works in this way, but some of the theoretical work produced implies a forced movement of the fish towards the light. Some of the work described above was carried out in water less clear than the Mediterranean and perhaps the aggregations were less pronounced. However, the relationship between aggregation and the degree of scattering has not yet been analyzed.

4. *Some specific quantities*

As the quantitative use of echo sounders increases, so does the search for particular values. The sizes of individual fish and their densities per unit volume can be estimated as described in an earlier chapter. Here some estimates of fish speeds, fish densities, and shoal sizes are described.

(a) *Speeds of fish*

It is difficult to measure the speeds of fish directly with an echo sounder for a variety of reasons. If a fish swims quickly along a diameter track across the sound cone at a given depth, its trace on the paper record is not so long as when it swims slowly. The track shortens with the distance of its mid-point from the acoustic axis, as a function of the directivity of the transducer. Hence an average track should be considered in relation to the mean angle for the size of fish. If the signals from the average chord or track are to be considered as being received from the mean angle, then the mean track is given by:

$$a = [4(r_2{}^2 - r_1{}^2)]^{\frac{1}{2}}$$

where a is the length of the track, r_1 is the minimum range and r_2 is the maximum range. Then the mean track as defined from the average ranges (minimum and maximum) is given by $\bar{a} = [4(\bar{r}_2{}^2 - \bar{r}_1{}^2)]^{\frac{1}{2}}$ in meters, the mean track can be divided by the ship's speed in m sec^{-1}, V, and multiplied by the number of transmissions per sec, p, (i.e. $\bar{a}p/V$ transmissions or pings). This quantity, the mean number of pings, varies with target strength and if it is assumed that the fish speeds are only a small fraction of the ship's then differences observed at the same range are differences in target strength. Cushing (1967) used such a method to show that, in layers of herring in the southern North Sea and off the Norwegian coast in winter, fish were distributed in depth by size; the bigger fish were near the bottom of the layer in the southern North Sea but in the middle of it off the Norwegian coast. If the sizes ranged from 23–33 cm, speeds might have ranged from 0.69–0.99 m sec^{-1} (i.e., 3 lengths per sec) and the differences in target strength, estimated from the mean number of pings were much greater than this. The mean track passes through the locus of θ_m on a conic section normal to a circular beam in such a way that half the track lies inside the locus and half outside. Then it can be shown that

$$\bar{a} = (Z/\cos \theta_{min}) - [(Z^2/3)(3 + 4 \tan^2 \theta_m - \tan^2 \theta_{min})]^{\frac{1}{2}} \tag{40}$$

where Z is depths,

From this form of equation the mean track can be calculated for any size of fish and depth; so as in Chapter 4 a distribution of fish sizes could be calculated from a series of measurements made on a paper record. Many paper records are in fact not suitable for this purpose, but a high speed Alden recorder can be used for research purposes.

Haram (1965) noticed that the mean number of pings on a coregonid (its local name in Lake Bala, in North Wales, is the Gwyniad, *Coregonus lavaretus* Linnaeus) at the same depth was five to ten times greater by night. A similar analysis of salmon traces in the open Pacific (Sano, 1968) suggested that the distribution of relative speeds correlated well with the catches of salmon in the surface gill nets. Because the fish were very much of the same order of size, such a method is reasonable, the higher catches being found when the fish were swimming faster, and when the traces appeared to be of shorter duration. Thus, with a little care, the traces of individual fish can be used to make estimates of their speed; of course, if the amplitudes were sampled directly at the same time to give some estimates of mean sizes, it is possible that estimates of absolute mean speeds would become available.

A fuller account of the estimation of the speed of fish with an echo sounder is given by Shibata (1968). He derives two simple equations for horizontal motion and for vertical motion:

$$\text{horizontal} \quad V_F = [(h_1^2 - h_2^2)/t]^{\frac{1}{2}} - V_s \tag{41}$$

$$\text{vertical} \quad V_F = [(h_1^2 - (V_s/t)^2 - h_2^2)/t]^2 \tag{42}$$

where V_F is the speed of the fish; V_s is the speed of the ship; and h_1, h_2 are perpendiculars on the paper record from the surface at the beginning and end of the time interval, t, during which the trace of an individual fish is recorded.

Shibata worked on tuna of about the same sizes and found that their speeds varied from 0.05–4.0 knots. This range of speeds suggests that the fish spent some time at attack speeds and some at a low cruising speed. From Shibata's distribution of speeds it is possible to deduce that 10–20% of the time was spent in attack and that the remainder was spent at a low cruising speed (< 3 lengths per sec as found in the laboratory for some much smaller fish Bainbridge, 1960). Saila and Flowers (1967) have suggested that because the chance of detection reaches a maximum at a low speed for the area that can be searched, a predator need not travel faster than this low speed. It is possible that the speeds measured under aquarium conditions are exaggerated a little, because feeding is separated from swimming; in the wild, the fish swim at the optimal speed to feed, but in the tank they may not be able to optimize their speed.

Thus, with care, there is no reason why the echo sounder's ability to discriminate single fish should not be used to make estimates of their average speeds. With the ship laid, the speed differences are minimized and the information, as number of pings, is increased. Then if average size can be specified independently (from the average amplitude), the distribution of number of pings indicates absolute speed.

(b) *Sizes of shoals*

Much of the description of echo traces in Chapter 2 was really concerned with size of shoal. The gross differences are those between "layers", "plumes", and "comets". A layer may extend for great distances but within one transmission the sound cone is filled for a prescribed vertical distance. Very often the most intense signals are received from the center of the layer, as if from a mean range, at the mean angle. Fish appear to be rather thinly distributed in layers and they are found at night in deep water or in turbid water in daytime.

But the dimensions can be readily given in horizontal distances and in thickness. In the early fifties the herring shoals extended at night for seventeen miles continuously north easterly from the Sandettié L.V. in the Straits of Dover. Aoyama (1960) has described three shoals of hairtails in the East China Sea; they were 4.0–5.5 miles long, by 30–40 miles wide and there was about 50 miles between shoals. They were about 10 m thick; Fig. 72 shows a diagram of such shoals reconstructed from echo sounder transects. Alverson (1967) has shown that the shoals of Pacific hake were 12 km in length along the depth contours, by 0.3–0.5 km in width and they were 6–20 m thick; the spawning shoals were thicker, about 40 m. The overwintering shoals of Norwegian herring in the East Icelandic current have been described by Truskanov and Scherbino (1966), (Figs. 65 and 66). Bolster (1958) charted the herring shoals on their spawning grounds in the Straits of Dover and found that they lay along the tide in long narrow bands; Harden Jones (1962) made use of this feature subsequently to make some estimate of their movements with the tide. It is interesting to recall that Belloc (1935), observing sardine shoals off the coast of Brittany from a balloon, found that the shoals of sardines were elongate with their long axes in the direction of the tidal streams. An echo sounder can be used effectively to gain substantial information on the sizes of large shoals which exist as layers in the sea. It is not yet known how such large shoals build up, how long they last, or how they vary from day to day, or during longer time periods.

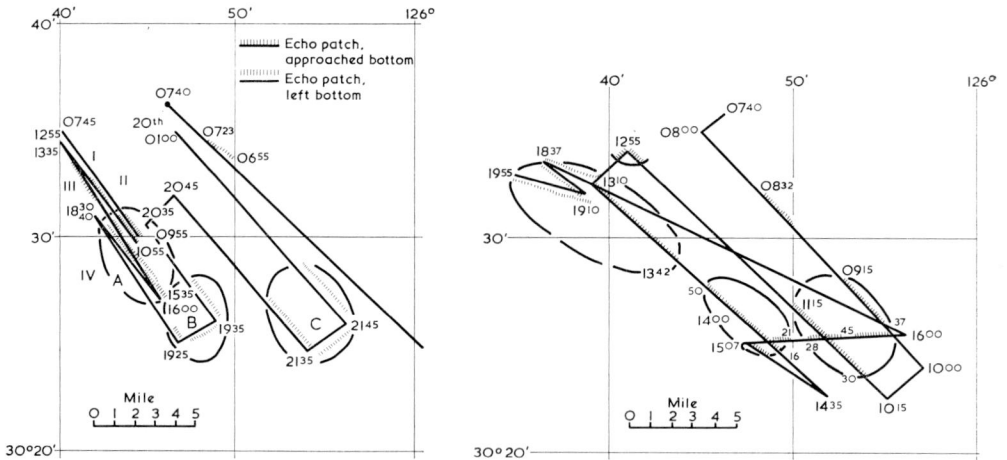

FIG. 72. A three-dimensional diagram of some shoals of hairtails in the East China Sea constructed from echo sounder transects (Aoyama, 1960).

A "plume" trace is made up of a number of transmissions, 20 or 30 on a shoal, which has considerable vertical extent in the water. Because the plume trace has a characteristic apex, indicating a shortening of range as the shoal crosses the sound cone, the shoal is smaller than the sound cone. But, as shown in Fig. 37b, there is a tail of reverberation in the lower part of the trace which indicates that the fish must be quite densely packed to generate it. If the top of the plume is at 30 m, the cone may be 20 m across, which means that the shoal is less than 20 m across. Such plume traces may be found in depths of up to 100 m; so the shoals are probably less than 20 or 50 m across according to depth. The vertical extent of such traces ranges from 10–30 m, which is a low estimate to take some account of the possible reverberation between fish recorded on the trace. Thus a shoal of average size is 20 m horizontally by 10 m vertically or 50 m horizontally by 30 m vertically in deeper water. It may be ellipsoidal in shape with the long axis in the horizontal dimension.

Similar measurements could be made on smaller shoals, the "comet" shoals of the paper record, which are probably those most frequently observed. However, each measurement is in range and not in depth, and for any particular shoal its angle from the acoustic axis and hence its range is unknown. Consequently, the problem must be approached statistically and the measurements averaged for each interval of range. As noted in Chapter 2 the thickness of such a trace at its apex is a good measure of its extent in range. So an average of such measurements in a short interval of range represents the range at the mean angle. But the measurement is one of differences in range so if the range interval is small, the bias due to differences in range or angle is small and the major cause of differences in the measurements is that due to size of shoal. Measurements of amplitude of received voltage, using a TVG, could be made at the same time; if packing density were constant for a given size of a fish species, the square of the voltage should increase with shoal size.

The study of shoal sizes has not proceeded beyond the gross stages described for two reasons; the first is navigational, in that only the largest shoals can be described adequately within the turning and position finding limits of a research vessel. The second reason is statistical, in that the position of a small shoal in the sound beam is indeterminate. However, with present day navigational aids and a statistical approach, a fuller study of shoal sizes by size of fish and species of fish should become available. In the last chapter of this book, a different approach will be discussed, the use of the ARL scanner to describe the shapes and sizes of fish shoals.

(c) *The packing density of single fish, or the distance between them*

From an echo record of individual fish, an estimate of their densities appears to be simple. However, from the statistical treatment set out in Chapter 4, the procedure is complex, requiring adequate averaging and sizing before any estimate of sampling volume can be reached. Strictly, the only estimates of density made following these procedures are those made by Cushing (1968b) on the South African hake populations. Even then, these estimates are only based on an average length of 46 cm and the length distribution, obtained by acoustic means alone, was biased, over-estimating the densities of larger fish and under-estimating those of smaller fish.

Cushing (1969) tabulated densities, distances between fish and sizes of fish for four groups of fish.

TABLE 7. DENSITIES AND DISTANCES BETWEEN FISH OF DIFFERENT SIZES

Groups	Length (cm)	Density ($n\ m^{-3}$)	Inter-fish distance	
			(m)	(lengths)
Sardine	20	2.0	0.79	4.0
Herring	30	0.7–1.0	1.00	3.3
Gadoids	70	0.00001–0.00008	43.8–90.9	60–130
Tuna	100	0.000001–0.0001	22.2–100.0	22–100

These figures were taken from Cushing (1957), on pilchards; Truskanov and Scherbino (1966), on herring; Cushing (1968) on hake; and Shibata (1962, 1963) and Nishimura (1961)

on tuna. So the distance between fish increases with the size of fish; but the bigger fish live at greater distances from each other than might be expected merely from linear differences in size; perhaps the increase is in proportion to their volumes.

(d) *Use of the specific quantities*

In particular regions, individual fish and shoals can be identified fairly well; for example, off Svolvaer, in the Vestfjord in northern Norway, in March, it is likely that most of the individual fish traces are from single cod. Similarly, shoals of fish in the Straits of Dover in November and December used to be composed almost exclusively of herring, as witnessed by the disappearance of traces with the disappearance of catches as that stock of herring declined. In contrast, there are areas where fish mix, as in most tropical seas and in many temperate seas and in such areas the difficulty of identifying the targets may preclude a proper use of the specific quantities.

Given this preliminary restraint of identification, echo sounders can be used to estimate speeds of fish, the densities of individuals, and the sizes of shoals. It is likely that each estimate is somewhat biased, but the differences between estimates are so much greater than the biases that they are worth making. Shibata's work on the observed apparent speeds of fish has suggested that tuna swim more slowly than might have been expected from laboratory experiments on much smaller fish. This is the sort of unexpected result which needs testing.

5. *The availability of fish to capture*

In this section the general problem which faces fishermen and fisheries biologists is considered: the dependence of catch on echo signal. Any correlation depends on how large is the fraction of vulnerable stock to invulnerable stock. Hence a direct study of catch per unit of effort, as stock density, is made when such correlations are attempted. The first indication of correlation appeared in the work of the late Skipper Balls, as described in Chapter 4. Recent work by Ogura (1968) has shown that four boats out of eleven fishing for mackerel out of Kawazu harbor increased their catches by 21% when echo sounders were used. A correlation was made by Cushing in Richardson *et al.* (1959) and it showed a dependence of catch of cod in the Barents Sea upon the signals recorded during the trawl haul. The signals were received predominantly from single fish (but not exclusively) and a correction was made for the inverse square spreading with increased depth. The relationship was curvilinear suggesting that at very high catches the trawl was too full to retain fish in the cod end; there were three observations of nil catch with high signal, when the signals were well above the bottom and probably above the headline. Burd (1964) used the weekly statistics for herring in the Southern Bight of the North Sea to correlate catches and echo trace (in mm per 10 miles) from echo surveys made during the East Anglian herring season; Figure 73 shows the clear correlation which was obtained. Thus there is a relationship between catch and signal, which must be well known to fishermen who buy echo sounders in such large numbers and so critically.

However, there is little evidence on how fish behave with respect to gear. Harden Jones (1956) showed quite convincingly that some herring shoals on their spawning grounds in the Straits of Dover moved with the tidal streams. So if the fish are to be caught by drift nets, which "drive" with wind and tide, there must be a differential movement between fish and net. Recently, Margetts (personal communication) has shown such a differential movement

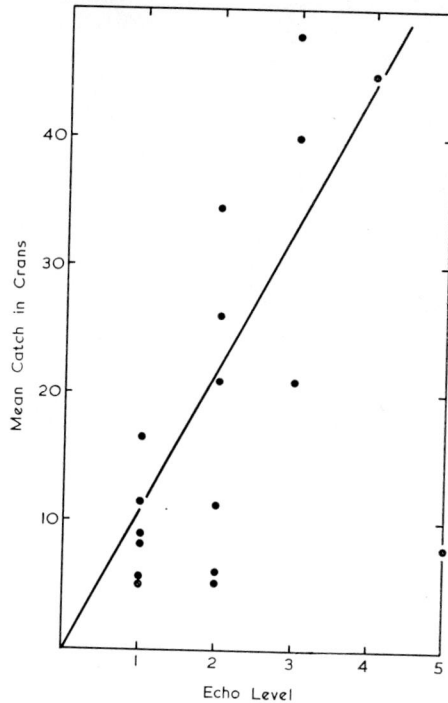

Fig. 73. Correlation between catch and signal (Burd, 1964); the catch of herring is expressed in crans (≡ 0.182 tons) and the quantity of echo trace in mm per 10 miles steamed in contemporaneous echo surveys.

by passing back and forth along a drifter's nets with a sonar firing sideways from the research vessel. Figure 74 shows the drift net as a long line, and the successive pictures show the herring moving towards and even into the net. These remarkable pictures show the whole process of fish capture taking place. Margetts (1967) has also shown that to catch herring with a mid-water trawl, in daytime, it is necessary to shoot it below the traces, whereas to catch pilchards, the fishermen should shoot at the middle of traces; hence, the escape speeds of the herring must be greater than those of the pilchard, which is not unexpected, because they are a bit bigger. Okonski (1969) compared catches of sprat and herring with the mouth opening of a mid-water trawl and concluded that a larger trawl catches more fish because the escape speed of the fish matters less. This is the basis of the work of Schärfe (see below) in his development of the very large mid-water trawls to catch herring. They are so large that the fish can only see one wing of the trawl in clear water and when they can see across it in the lower part of the trawl, it is going too fast for them to swim out of it.

Much of the success of the development of the German mid-water herring trawl originated from the netzsonde, a headline transducer. The first headline transducer was developed by Pye Marine in Lowestoft with a special torque controlled winch and this equipment and developments of it are used on English research vessels. But the most important development was carried out in Germany (Schärfe, 1960; v. Brandt, 1960; Steinberg, 1960, 1967; Mohr, 1963; Schärfe and Steinberg, 1963). Figure 75 shows two uses of the netzsonde; (a) to catch herring in mid-water, and (b) to catch fish, including herring very close to the bottom. Each shows the record from the ship's echo sounder and from the headline transducer and the

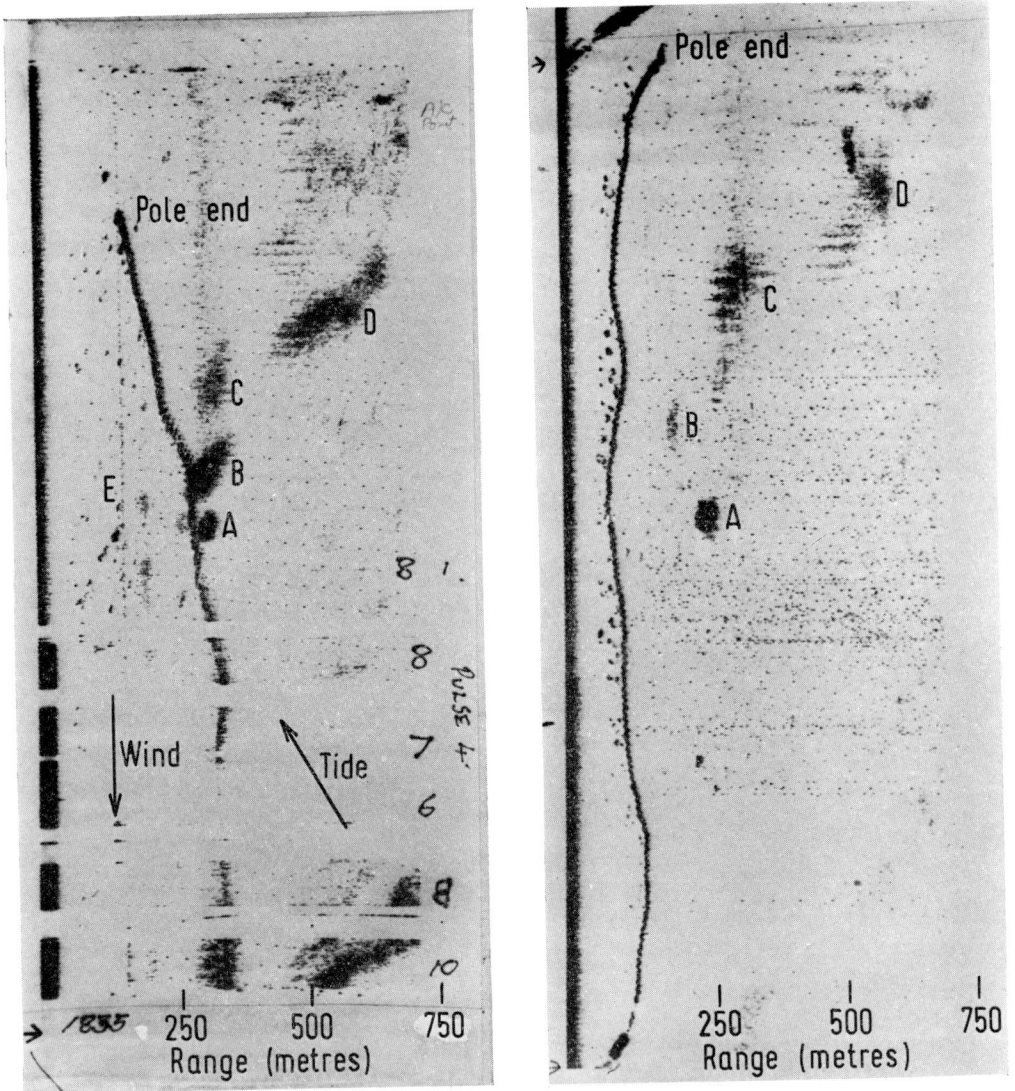

FIG. 74. A drift net, with shoals of herring approaching it, observed with sonar (Margetts, 1969, personal communication).

fish appear to remain at much the same depth as long as the trawl is large enough. The technique has been applied to single boat mid-water trawls on cutters, luggers, trawlers, and large stern trawlers. Presumably it is successful on the smaller boats when the water is rather turbid, so a smaller trawl can be used; of course such a trawl is still very large with a mouth opening in tens of meters. Alverson (1967) has used an analogous technique with mid-water trawls on the Pacific hake; the Cobb trawl is a large mid-water trawl with a headline transducer which skims above the bottom. The fish are caught in rather shallow water, about 70 m, but perhaps the light intensity is low and the fish do not see all the trawl because they do not dodge it as it is hauled rather slowly at about 1 knot.

Recently a top trawler skipper, Captain Charles Drever, and an echo sounder engineer, Mr. G. H. Ellis, have published an account of their experiences in fishing for cod in the

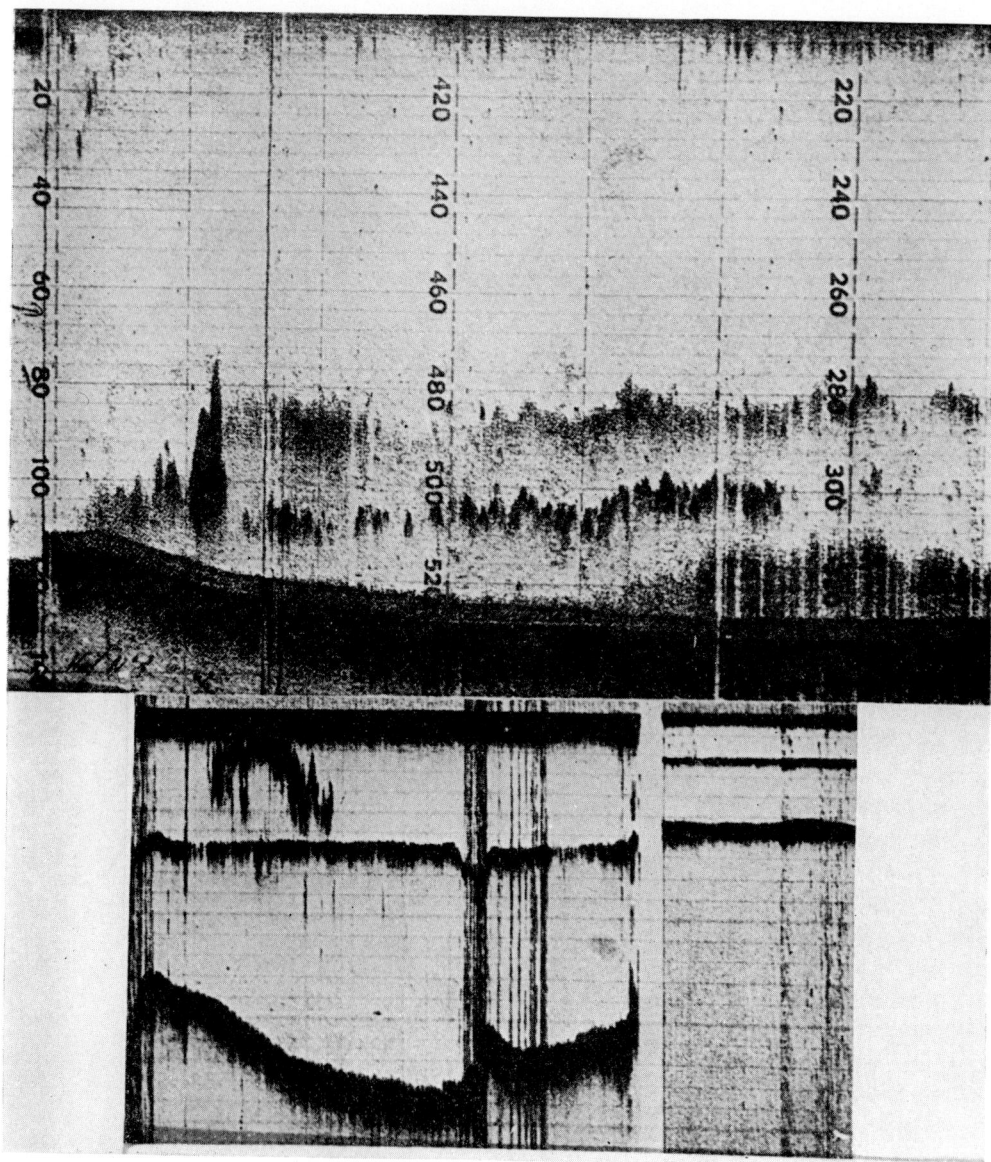

FIG. 75a. Shipborne traces of herring and netzsonde traces, in mid-water (Steinberg, 1967).

FIG. 75b. Shipborne traces of herring and netzsonde traces, near the bottom (Steinberg, 1967).

N.W. Atlantic with the Kelvin Hughes Humber Gear (Drever and Ellis, 1969). They found by experience that they could distinguish between signals from sandeels, cod, and redfish; those from sandeels were rounded on the CRT presentation and those from redfish were weak, as compared with the strong spikes from cod. They compared two grounds, Woolfall on the Grand Banks and Nanortalik off W. Greenland. Both were daylight grounds; on Woolfall, there were small groups of fish widely spread with single fish between them. The catch rate was steady, each haul taking 1–2 hr. On Nanortalik, there were large clumps of fish up to 5 fm off the bottom with no single fish between them; here the duration of tow depended on the distance between clumps and up to 400 baskets might be caught in 20 min from a single clump. Further, it appeared that the fish were moving because a second haul made on the expected position of a clump sometimes yielded nothing. Working on Woolfall Bank, Captain Drever found that the record of the first haul could be used as a standard for the whole day and that during this day catches could be reasonably predicted by comparing the records with that of the first haul. In general they found that a correlation of catch and signal was only valid if the echoes recorded within 60 cm of the seabed on the triggered pen recorder chart were used (Fig. 76).

One of the most interesting points about this account of echo fishing is the deliberate adjustment of the gear according to the vertical distribution of the fish. If fish were randomly distributed up to the headline, then the latter was floated up to catch more; if they were hard down, the floats were taken off to obtain maximum spread; and for the same purpose the bobbin size was reduced. Drever and Ellis point out that when towing along a contour because of the slope error, fish are lost to the echo sounder on the deeper side. Further, because the area sampled increases with depth, the masked area, or dead ground in which fish are hidden, is more extensive; although as shown in Chapter 4 at still greater depths, for a given size of fish such areas will tend to shrink as the sampling volume decreases. Thus the whole system is continuously modified on the basis of information from the echo sounder and from the catches. Indeed, partly as a result of their experiences, the White Fish Authority, in the United Kingdom, developed a new, but expensive, transducer, with a steered narrow beam; the transmitter has a 3.5° beam athwartships and 9° in the fore and aft plane and the receiver a 4.5° beam fore and aft and a wide beam athwartships. Thus the effective beam was considerably narrower than that of the original Humber gear. The beam is steered electronically to receive signals only when the transducer is normal to the seabed, so eliminating the effect of rolling. Not only does this make the whole system more powerful and allows discrimination to greater depths, but the resolution of single fish in time, from transmission to transmission, is increased (Fig. 77).

This very brief account of the work of Drever and Ellis only highlights the study of behavior made by skipper and engineer as they worked at sea together over a period of years. For details of the techniques developed, particularly with Decca track plotters and Kelvin Hughes warp load meters, their original papers should be consulted. The most important result for our present purpose is that they have arrived at the same conclusion as the fisheries biologists, reached in the last chapter—that a proper estimate of the volume sampled is needed before a catch and signal correlation can be expected to yield as clear results as those of Dowd with the transducer towed at a fixed height above the seabed.

With the exception of Dowd's work on discrete catches, and that of Burd, which summerizes all the information from a fishery, the correlations between catch and signal obtained by fisheries biologists have only been sporadically successful. On the other hand, Drever and Ellis are sufficiently confident to rely on their echo sounder to predict catches at least

FIG. 76. Cod traces "hard down" on the bottom on a triggered pen recorder chart from a Kelvin Hughes Humber gear (Drever and Ellis, 1969).

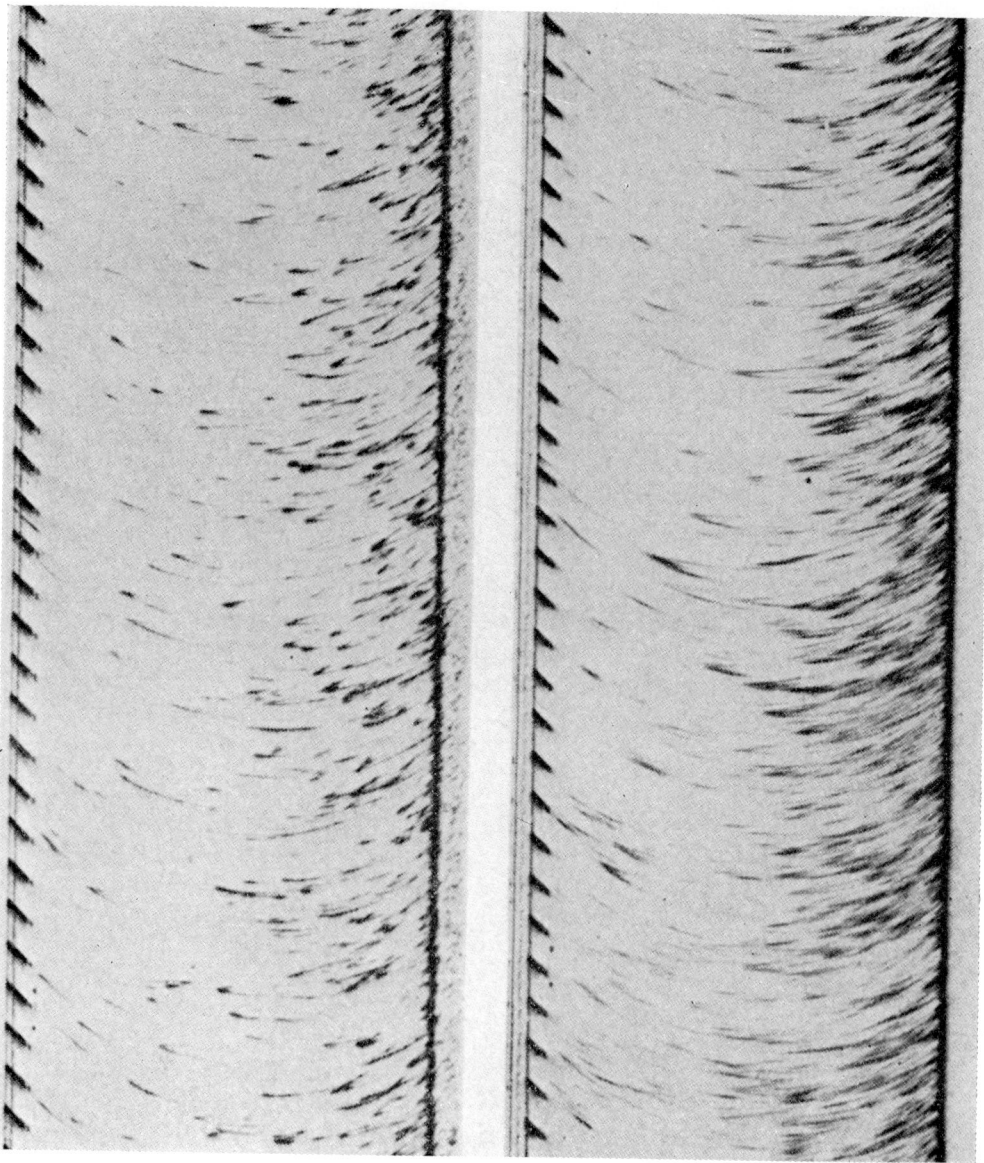

FIG. 77. Comparison of traces from individual fish, recorded with an ordinary Humber gear and with the steered narrow beam equipment (Drever and Ellis, 1969).

on a day to day basis and within 60 cm of the seabed. Hence with the volume corrections properly applied which deal with this particular point, it is likely that such correlations should become successful. Hence availability can be investigated directly and in detail. Up to the present it has been investigated in an inferential manner from stock density data or from variations in mortality coefficients.

6. *Conclusion*

The study of fish behavior is an essential part of the discipline of a fisheries biologist, partly as the general biology of a fish species, in habits, in migration circuit, and so on, and partly as the response of the animals to gear. The investigation of diurnal rhythms revealed the variations in vertical migration and in shoaling behavior; of great interest is the link between the two phenomena in the responses of fish to light intensity. The effect of artificial light on fish is of less general interest, but possibly depends upon a relationship which has not been sufficiently investigated, the forced movement of clupeids towards a light source in darkness, and to what extent scattering limits the forced movement. It was shown that the echo sounder can be used, in a rough manner, to make estimates of fish speeds, shoal sizes, and densities of individual fish and it was suggested that tuna in the open sea swim more slowly than smaller fish in tanks, that spawning shoals are bigger than feeding shoals, and that, as individuals, bigger fish live further from each other than little fish. The quantities are in themselves interesting and perhaps have been under-exploited when echo sounders are used at sea.

The most valuable aspect of fish behavior is that in relation to fishing gear, but it is the least developed. Recently with drift nets and with mid-water trawls it has been shown that the mechanisms of capture could be described in detail. But for the more complex case of the bottom trawl, the work is incomplete because the correlations of catch and signal have not been attempted taking into account the differences in sampling volume which are needed. Further possibilities are discussed in a later chapter on new acoustic instruments.

CHAPTER 6

THE DEEP SCATTERING LAYER

1. *Its discovery and early history*

AN EXAMINATION of fish detection would not be complete without a brief study of the Deep Scattering Layer. It is a layer recorded by echo sounders in the deep ocean, discovered by American sonar experts during World War II. Duvall and Christensen (1946), Eyring *et al.* (1948), and Raitt (1948) described the wartime results, which included records from echo sounders by day and night, over the deep ocean, and some oscilloscope measurements of amplitude. Uda (1956) has noted that the N-layer discovered in deep water during World War II off Nojuna Cape in Japan corresponded to the Deep Scattering Layer in the Atlantic. Moderately powerful echo sounders, with long pulse lengths (*ca.* 0.1 sec) were used. So the layer extended in depth on the echo record for 0.1 sec (or 75 m in echo distance) and, as the signal-to-noise ratio increases with longer pulse length, the equipment was quite sensitive. The records were extensive in distance and depth. They were perhaps recorded at a lower level of sensitivity than that used today by fishermen and fisheries biologists using the short pulse length echo sounders. The layer has been recorded in all the oceans and it extends for hundreds of miles (Fig. 29) and rises towards the surface at night. However, with a short pulse length (*ca.* 0.001 sec) the vertical extent of the layer is much the same as that recorded with the long pulse length, as shown by Kocsy (1954) on the Albatross expedition. Figure 78 shows a typical record of the oceanic layer migrating towards the surface.

The biological interpretation of the vertical migration of the layer was started by Johnson (1948), who pointed out that the phenomenon was characteristic of animals living in the deep ocean, possibly euphausids, which certainly migrated over such an extensive depth range and like the layer are present in most of the world's oceans. Hersey and Moore (1948) summarized the results of investigations at a frequency of 18 kHz between Cape Hatteras and the Gulf of Mexico. The vertical migration ranged from 225 and 300 m to the surface and it followed roughly calculated isolumes based on extinction coefficients observed much nearer the surface. Moore (1950) pointed out that euphausids lived and migrated at the right depth ranges, but that sergestid and acanthephyrid crustaceans did not. As will be shown below, the latter conclusion may have been a consequence of the frequency chosen. Moore showed that a layer existed in the Mediterranean, south of the Balearic Islands, contrary to earlier assertions. He tried to correlate the evening ascent of the layer and its morning descent with calculated isolumes, but the fit deviated by many log units; however, the rate of ascent of 2–4 m min^{-1} was not considered excessive for euphausids. Further, euphausids were caught in plankton nets shot in the layer. But Chapman (1947), an ichthyologist, thought that the layer was composed of oceanic fish, louvars (or coryphaenids).

Dietz (1948) examined long series of records and found that the layer existed in all oceans, except the Antarctic; Tchernia (1949–50) confirmed this observation, but in

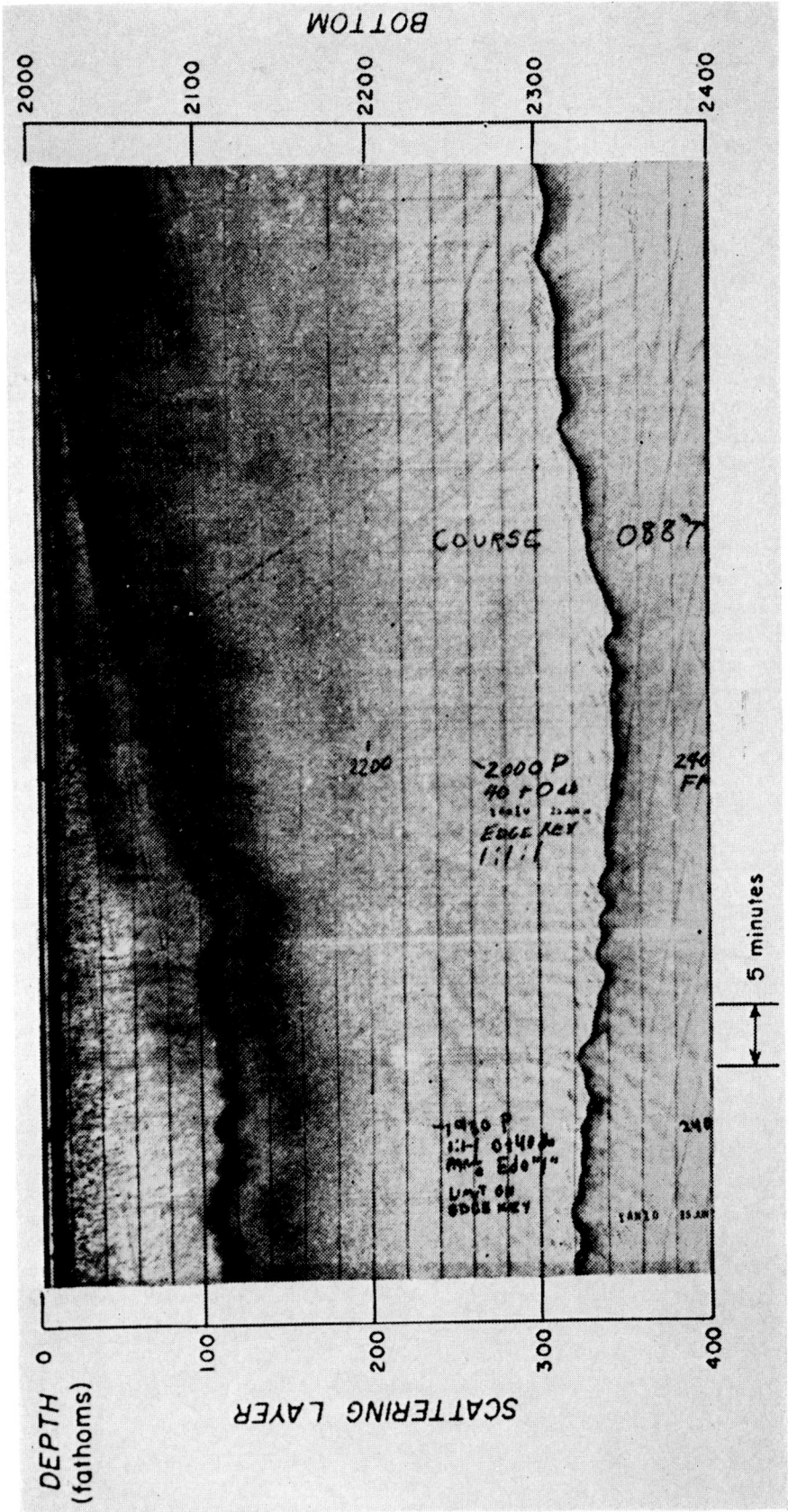

FIG. 78. The Deep Scattering Layer as recorded in the early days of the investigation (Hersey and Backus, 1962).

Tchernia (1951) he discovered a layer in the Antarctic (at 24.5 kHz) in December, January, and February, at a depth of 360–630 m which did not migrate vertically very much. The Antarctic contains the densest concentrations of euphausids in the world ocean (which would be expected to migrate to the surface at night) and so Tchernia's observations at the time were critical. However, it is likely that the layers in the Antarctic were much less intense than those in other oceans. Recently, Marshall (1971) has noted that mesopelagic fish with swim bladders are absent from extreme polar waters.

The most important paper on the Deep Scattering Layer in the early stages of investigation was that by Marshall (1951). He accepted that the layer was comprised of animals, but noticed that they should be large enough to scatter sound back to the receiver, that they lived in all oceans save the Antarctic, (Tchernia's second paper had not then been published), and that they migrated from 150–400 fm by day to the surface by night. Raitt (1948) had suggested from volume reverberation measurements that the scatterers were resonant gas bubbles, about 1.5–2.5 mm in diameter at a density of about 10^{-4} m^{-3}; the cross sections of individual scatterers ranged from 0.1–10 cm^3. From these facts, Marshall concluded that the scatterers were air-bladdered plankton eating fish, myctophids, sternoptychids, or *Cyclothone*. Marshall (1971) shows that 80% of migratory species had swim bladders. There are few bathypelagic fish in the Antarctic (where the layer was not prominent) and their rate of ascent or descent would be limited by pressure changes on the air bladder. Lastly, the bathypelagic fish are not very large and their air bladders correspond in size to that suggested by Raitt from the scattering cross sections.

Tucker (1951) examined catches from two layers in the San Diego Trough made with a net superficially resembling the later Isaacs Kidd net. From a shallow layer, which scattered sound less intensely, between 100 and 200 m in daytime, euphausids were caught. But in the deeper layer, between 300 and 500 m in daytime, myctophids were caught with a few squid, not enough sergestids, and acanthephyrids, and very few euphausids. The fish were caught at a density of 0.034 per m^3, which was a greater density than that observed acoustically by Raitt.

This brief history of the early investigations of the Deep Scattering Layer started with its discovery during World War II, and subsequently it was found in all the world's oceans, even to some extent in the Antarctic. The solution arrived at by Raitt, Marshall and Tucker was the right one, but much additional work was needed before the solution was confirmed by independent methods, physical and biological.

2. The physical investigations

The physical investigations may be grouped in three parts: (a) the analysis of records taken from a transducer suspended in the layer, (b) frequency analysis of the layers, and (c) the determination of volume back scattering by single scatterers which was an extension of the work started by Raitt (1948).

A transducer operating at 12 kHz was suspended in a layer off New England (Kanwisher and Volkmann, 1955) and single scatterers, i.e., as fingernail traces on a paper record, were recorded at a density of one in 650 m^3. Johnson *et al.* (1956) continued the observations off Puerto Rico. Figure 79 shows a typical record of single scatterers as the layer was rising. The transducer sampled a volume of about 13,000 m^3, and that occupied by a single scatterer was about 650 m^3. Off New England, the observations were repeated and a brass ball, 6.17 cm in diameter, was suspended below the transducer, partly as a calibration device,

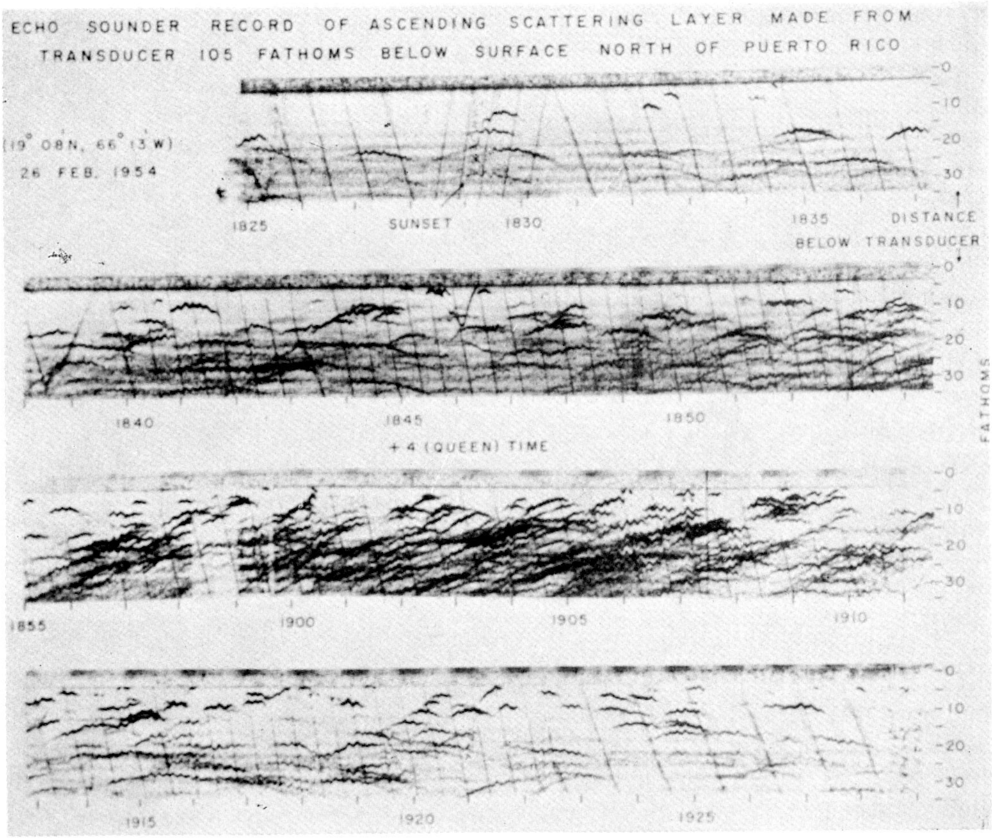

FIG. 79. A record of single scatterers within the layer (Johnson *et al.*, 1956).

but also to obtain estimates of target strength or scattering cross section of the targets; the target strength of the ball was −38 dB at 12 kHz and many scatterers gave the same signal and a few yielded a larger signal. Very roughly, from our present knowledge, the scatterers would have been fish < 10 cm in length. The following table summarizes the density observations by depth:

TABLE 8. VOLUME (m³) PER SCATTERER
BY DEPTH

Depth (fm)	m³ per scatterer
39.0– 51.5	850
68.0– 85.5	1500
107.0–119.5	1800
139.0–151.5	1800
178.0–190.5	2100
212.0–224.5	820
250.0–262.5	1200
285.0–297.5	1050

A camera was attached to the system and it was found that all pictures of fish were accompanied by strong echoes; one photograph was identified as *Nealotus tripes* Johnson, of which only eight specimens had been previously recorded.

The frequency analysis of the Deep Scattering Layer was started by Hersey *et al.* (1952) and continued in Hersey and Backus (1954), and Hersey *et al.*, (1961); the theoretical development, considering the volume scattering coefficients at different frequencies and ranges, is given in Machlup and Hersey (1955). A small charge was detonated in shallow water and signals were received on a wide band hydrophone and were fed through filters into channels of specified frequencies or directly on to a sound spectrograph. Frequency dependent layers were located at 3.5, 15, and 20 kHz, and the low frequency layer lived at deeper levels in the daytime than the high frequency one. During the vertical migration of the layer, frequency decreased with ascent and increased with descent. Figure 80 shows a frequency analysis of the layer during ascent; at 1846 hr the three components are visible at three distinct ranges of depth. The layer at 200–250 m responds at 20–26 kHz; that at 250–300 m responds at 8–15 kHz, and there is another layer extending over a considerably vertical extent; after ascent has occurred, the shallow layer dropped in frequency at 16 kHz, the middle layer to 6 kHz, and the lower layer, which did not migrate, did not shift in frequency at all. The deeper non-migratory layer was probably resonating at 3–4 kHz, which suggests that they were fish rather larger than the myctophids, perhaps of sardine size. The frequency dependences on depth for the other two layers followed separate relationships; in the shallow layer, frequency declined during ascent as the five-sixths' power of the depth and in the middle layer it declined as the square root of the depth. If the swim bladder of the fish remains at constant volume during ascent and descent (see Chapter 3), frequency should vary as the square root of the pressure difference, or as the square root of the depth. If it expands and compresses as a function of pressure during ascent and descent, then the radius of the swim bladder varies inversely as the cube root of the pressure and so frequency should vary as the five-sixths power of the pressure difference, or depth. Hence, it would appear that the upper layer consisted of fish which allowed their swim bladders to expand and contract during migration; this implies there must be an upper limit to their capacity for migration, in expansion of the swim bladder, fairly near the surface. The middle layer apparently comprised fish which retained the swim bladders at constant volume during migration; on the face of it, they have an unlimited capacity for vertical migration, perhaps even to the surface itself, but it implies that all the gas available during potential expansion is absorbed by the retia mirabilia of the swim bladder into the blood stream. An alternative explanation is that the swim bladders are in fact oil filled, as in the case in some bathypelagic fish (Marshall, 1951). However, Marshall (1961) has shown that the retia mirabilia of bathypelagic fish are well developed; indeed their depth ranges are correlated with the areas of the retia. The fish do not rise very quickly, about 2–4 m min^{-1}, and Marshall believes that the large retia and the large blood vessels supporting them may allow the swim bladder to resorb oxygen into the blood stream as the fish rises in vertical migration.

Machlup and Hersey (1955) studied the volume back scattering coefficient at different depths, at one frequency (15 kHz). The coefficient expresses the total energy scattered back to the receiver from a range shell specified at a given depth as described in Chapter 3. They showed that the coefficient from the center of the layer was more than 20 dB greater than that shallower or deeper. Andreeva and Chindonova (1964) have used an analogous approach, but define their back scattering coefficient, m, as follows: $m_0 = N\sigma/4\pi$, where N is the concentration of scatterers and σ is the scattering cross section. They suggest that

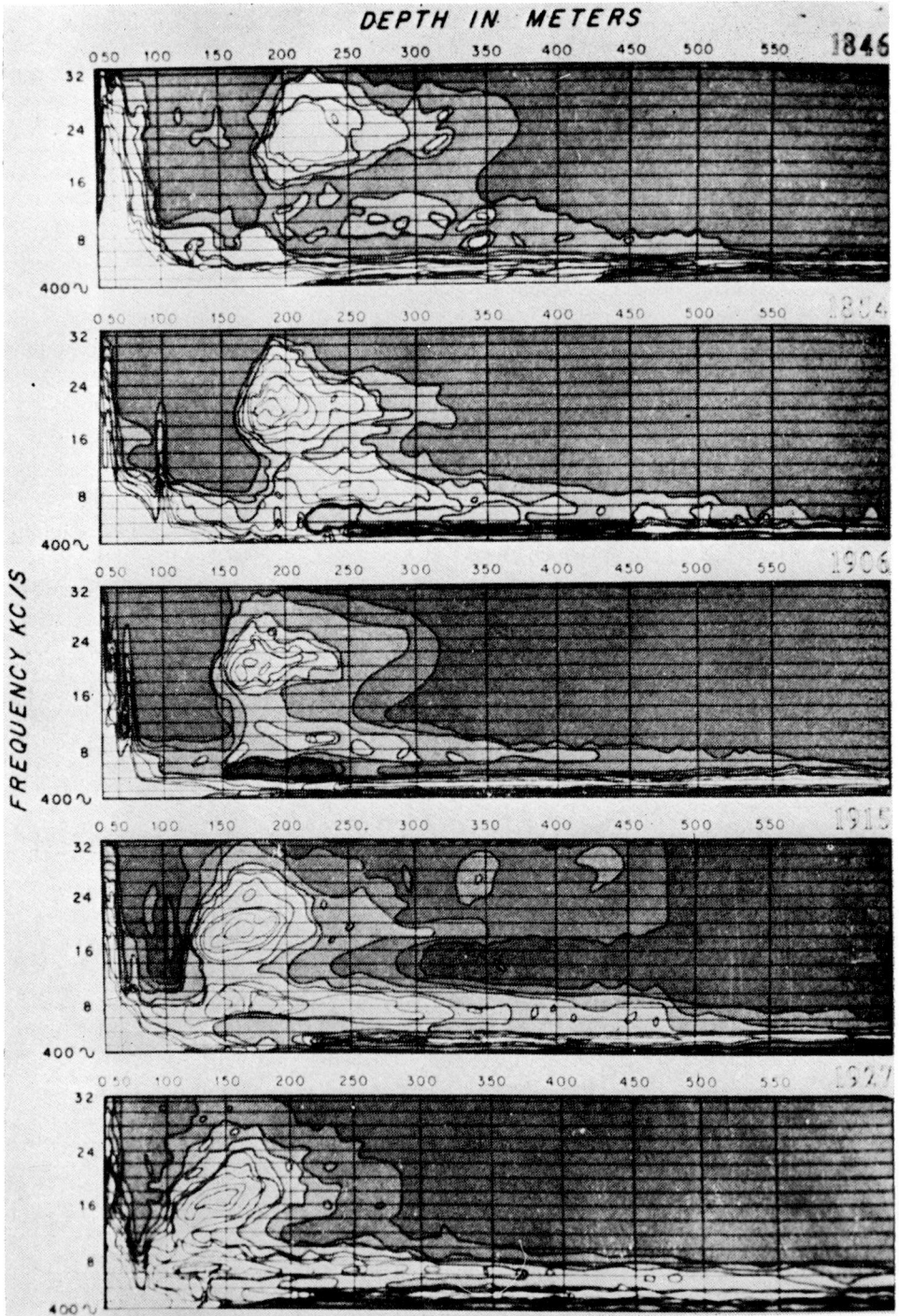

FIG. 80. Frequency analysis of a layer rising towards the surface at dusk (Hersey *et al.*, 1962).

there is a quality factor in air bladder resonance which varies with depth, being maximal at about the day depth of the layer. From the expected distributions of animals in the sea, a theoretical volume scattering coefficient can be calculated; Andreeva (1965) has calculated the trend of such coefficients with depth and has checked the distributions by sampling with a mid-water trawl. The method of Machlup and Hersey analyzes the distribution as received by the transducer only. That of Andreeva presupposes that the distribution of animals is given correctly by size in the net samples and that their scattering cross sections are estimated correctly. A useful extension would be to combine the two methods to detect failures in sampling or in estimation of scattering cross sections.

From the physical investigations conducted in the main by Hersey's group in Wood's Hole, Massachusetts, U.S.A., the layer has been virtually identified in its main constituents as being composed of bathypelagic fish, which are small from their scattering cross sections, and which are rather thinly distributed, from the sampling volumes of the acoustic equipment. The methods used are rather complex and probably have not been fully exploited. For example, Andreeva and Zhitkovski (1968) have detected layers with an explosive source and receiver filters at 7, 10, 15 and 20 kHz, at depths between 1000 and 2500 m. This suggests that the bathypelagic fish are as small as present fishing suggests. Recently, Blaxter and Currie (1967) have analyzed the vertical migration of layers off the Canary Islands with transducers working at 10, 36 and 67 kHz; Fig. 81 shows the ascent for 2 hr as recorded at the three frequencies. It is likely that each transducer selects its part of the migratory spectrum and so the apparent ascent might well be under-estimated. But given this caveat

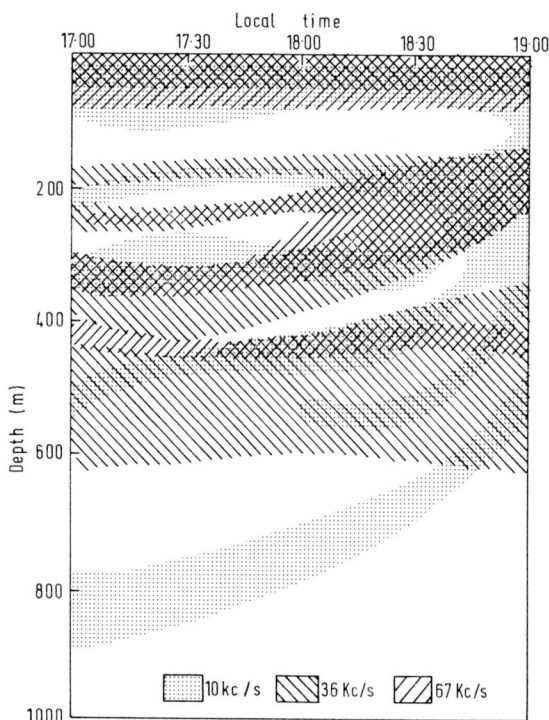

FIG. 81. The ascent of a layer for 2 hr as recorded at three frequencies, 10, 36 and 67 kHz (Blaxter and Currie, 1967).

the picture remains complex and there are four layers recorded at 67 kHz, a frequency which had not been investigated previously. It is likely that acoustic equipment for the examination of oceanic layers will become more complex and even more productive in the future.

Davis (1971) describes the volume scattering as a function of depth as recorded from airborne "dunked" transducers and a quasi synoptic picture emerges with diurnal differences of 10 or 15 dB according to area and varying with frequency. A similar study by Chapman *et al.* (1971) suggests that the density of scatterers off Newfoundland reached 10×10^{-6} m^{-3}. A more advanced technique has been developed by van Schuyler (1971) with a narrow band spectrum analysis (between 0.1 and 20 kHz) in range; given a relationship between size of animal and resonant frequency the size distribution and density of animals can be estimated. For biological purposes, no estimate of volume scattering is any longer of much value unless processed by something like van Schuyler's methods.

3. *The biological investigations*

The biological investigations since the early fifties may be grouped under four headings: (a) distribution, (b) animals caught or not caught in the layers, (c) animals comprising the layer, and (d) response of the layer to measurements of light intensity.

Tchernia (1951) has observed an echo zone for a brief period in the Antarctic, but it is clear from his report that the layer is not so strong or so widespread there as it is in other oceans. It was not until 1965, when Hunking reported a Deep Scattering Layer from Fletcher's Ice Island, that the Arctic Ocean itself has been properly examined. With a 12 kHz echo sounder a layer was found between 50 and 200 m, with some discrete echoes within it between June and August; in the following year it was found to persist from June to November. So the layer is found in all oceans, but its presence in the Arctic and Antarctic is less pronounced than elsewhere.

Beklemishev (1964) studied the distribution of scattering layers in the Pacific, including the Pacific Antarctic. He divided the whole ocean into biogeographical regions, according to the distributions of phytoplankton, zooplankton, and fish. He found that the migratory layers were not found in the Pacific Antarctic, but that there were such layers in the sub-Arctic. The non-migratory layers appeared to be distributed in depth in the same way both in the tropics and in high latitudes. An interesting point is that the macroplankton is concentrated in depth within the depth range of the main seasonal thermocline. The thicker the layer occupied by the main seasonal thermocline, the greater the vertical migration of the scattering layers and this migration reached greater day depths in tropical waters. But the most important point noticed by Beklemishev was that the layers tended to be more intense in the more productive regions. So we would expect intense scattering layers to be found in the eastern boundary currents, in the equatorial current system, and off the upwelling areas in general.

Various attempts have been made to capture the animals in the layers. Komaki and Matsuye (1956) examined the vertical distribution of plankton animals with reference to the scattering layers and showed that they did not live at the same depths. Hashimoto and Maniwa (1956), using frequencies of 14.5 and 24 kHz, examined layers off Yaiza, in Japan, and concluded that one was composed of *Sergestes* (on a ground where these animals are fished commercially) and that another was composed of yellowtails; perhaps such layers are special forms, close to the Japanese coast. Kinzer (1969) studied the distribution of zooplankton in relation to deep scattering layers sampled at a frequency of 15 kHz off

Portugal, the Canary Islands, and off W. Pakistan. He used a modified Gulf III sampler which he towed at 5–6 knots and so it is unlikely that the macroplankton animals escaped from its path. It was shown that euphausiids and copepods were more abundant, in numbers and in displacement volume, at the depth of the Deep Scattering Layer than in the water above it or below it. The average density of euphausids was 0.3 per m^3 in the eastern Atlantic; Hersey and Backus (1962) suggested that a density of 200 per m^3 was needed to generate the observed scattering intensities. Shibata (1963) towed an Isaacs-Kidd trawl through a layer at 400–700 m in daytime and at the surface at night. Euphausids, molluscs, siphonophores, sagittae, and myctophids were caught; a similar array was found in the guts of tuna, which means that the Deep Scattering Layer comprises an intermediate trophic level between tuna and the macroplankton. Hence, the layer, in general, consists of macro-plankton, including euphausids, and of fish like myctophids feeding on them. The euphau-sids are often caught in the layer but the signals are received from the fish.

Blackburn (1956) had suggested that shallow scattering layers were generated by hetero-pods, with gas bladders. Hansen and Dunbar (1971) have shown that signals are recorded also from pteropods. Barham (1963) has seen siphonophores in the Deep Scattering Layer from the bathyscaphe *Trieste*, together with myctophids, sergestid shrimps, and immature hake (*Merluccius productus*). The siphonophores were identified as *Nanomia bijuga* Delle Chiaje and their nectophore gas bubbles in preserved material were 1.49 mm in length by 0.97 mm in breadth, so they would resonate to 12 kHz at about the day depths of the layer. Populations as dense as 0.3 per m^3 were observed. The nectophores have gas glands and a pore through which gas can be excreted and the gas is mainly composed of carbon monoxide (Pickwell *et al.*, 1964); it was observed that the gas gland can produce bubbles fairly steadily and so the observed migration in ascent of 300 m h^{-1} is not unreasonable. Barham (1966) examined the proportions of siphonophores and myctophids in a layer off Cape San Lucas in Bahia, California. In general both groups were seen within the layer in numbers and only odd animals were observed outside it. The work was carried out at night and the animals were observed by lights switched on for short periods (of about 2 min); as the layer extended above and below the bathyscaphe, up to 100 m in both directions, it is unlikely that the results were very much biased by the lights. During two dives at night 394 myctophids were seen in the layers and 11 outside them; 104 physonects were seen in the layers and 14 outside them. Thus both groups of animals were very much more abundant within the layers. From further measurements of the bubbles (1.9–3.4 mm diameter as spheres) Barham suggests that resonance would occur between 15 and 25 kHz. Kinzer (1969) caught nectophores of *Agalma okenii* Eschscholtz and *Hippopodius hippopius* Forskål at a maximum density of 1 per m^3 in layers. Such densities are much higher than those observed acoustically by Raitt (1948) and Hersey and Backus (1954); perhaps the proportions of animals with nectophores of the right size (i.e., large ones) are rather small. Then the whole populations would be best sampled at rather higher frequencies and the acoustic equipments for sampling the Deep Scattering Layer have selected the bigger animals. It is not yet known whether the siphonophores are widely distributed enough at a sufficient density to be identified as a general cause of the layer. But in three positions, at least, they are.

One of the most interesting biological problems in the study of the Deep Scattering Layer is its dependence upon light intensity. In the early period of examination, the isolumes were calculated and the correlation between them and the vertical migration of the layers was only a rough one. With the application of the photomultiplier to underwater light studies, it became possible to measure daylight intensities at the day depths of the layers and

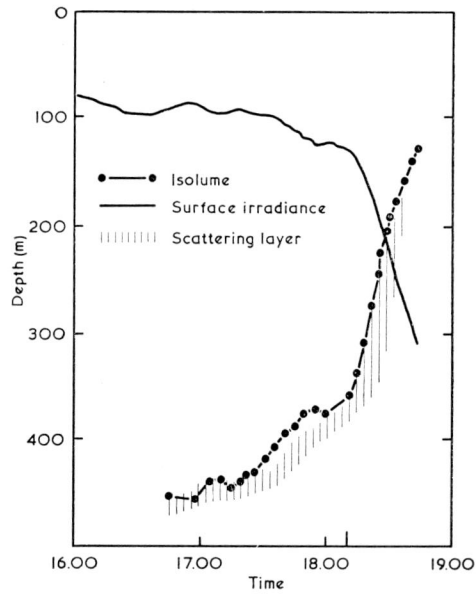

(a)

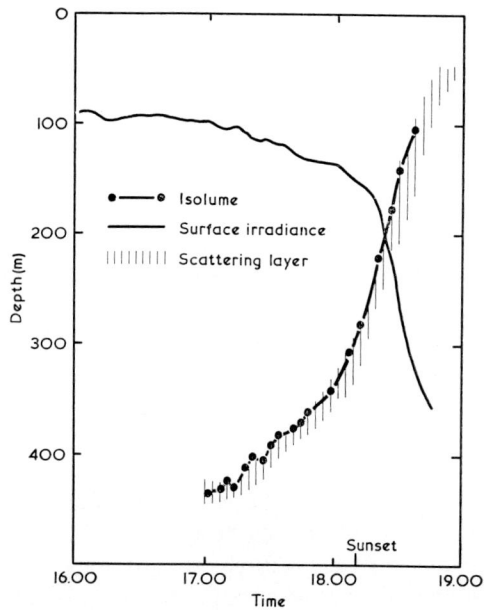

(b)

FIG. 82. Correlation between isolumes and depth of a Deep Scattering Layer in ascent and descent (Boden and Kampa, 1967).

bioluminescence both day and night (Clarke and Wertheim, 1956). Kampa and Boden (1954) using a 17.5 kHz transducer, showed that a layer rose with the 6.6×10^{-4} foot-candle isolume and descended with that at 5.9×10^{-5} foot candles, which suggested that the animals had become dark-adapted during their night near the surface. In the Golfe de Lion in the Mediterranean, it was found that, in ascent, the layer "overtook" the estimated depths of the isolume, when there was much luminescence (Boden and Kampa, 1957). A later study by Boden and Kampa (1967) showed that the dusk ascent and the dawn descent were closely correlated with the isolume 5×10^{-4} mcW cm^{-2} (with a filter centered on 474 nm, Fig. 82). In general, the animals lived at $3.5–7.5 \times 10^{-4}$ mcW cm^{-2} and there was no dark adaptation. Boden *et al.* (1960) believe that the light reaching the Deep Scattering Layer is modified by bioluminescence at 478 nm, which is the wavelength at which three euphausid species transmit. The narrow band filter excluded the light from sources other than bioluminescent organisms. Bioluminescent sources can flash at a frequency of as much as 160 per min at 100 m and can reach intensities of 10×10^{-4} mcW cm^{-2}, which is brighter than the intensity observed at the isolume (Clarke and Denton, 1962). The parts played by bioluminescence and the light from the surface have not yet been separated fully during the course of a diurnal migration. At the day depth, bioluminescence might be a predominant source, but during ascent and descent this is unlikely if only because all animals transmitting and receiving bioluminescence would then be expected to migrate at the same speeds. Perhaps during this period the two groups of animals tend to separate and the major part of the migration itself is mediated by the surface light intensity.

Recently, Dickson (1972) has studied the depths of scattering layers from the Navado Atlantic surveys recorded at 10 kHz between 10 and 68°N. Two layers were recognized as being of a widespread occurrence, one at a day depth of 215–280 fm, found south of 46°N and another all over the ocean, at a day depth of 175–330 fm. Dickson assembled a large quantity of data on oceanic transparency from Secchi Disc observations. The day depths of both layers were shown to be correlated well with the estimates of transparency, which perhaps implies that the day depths of bioluminescent organisms are themselves correlated with the light intensity from the surface. Thus, the precise measurements made by Boden and Kampa are confirmed on a very broad scale indeed.

The biological investigations have supported the physical ones. As might be expected the oceanic distribution of the layers coincides with biogeographical divisions and the most intense layers are found in the most productive regions. Most animals have been caught in the layers, primarily fish and euphausids, but siphonophores and sergestids have been associated with some particular layers. But it is unlikely that the general thesis that the layer is composed of myctophids can now be denied although siphonophores contribute to it in certain regions. Although modified by bioluminescence, the migration of the layer of little fish probably occurs within a fixed isolume during ascent and descent, which recalls Russell's thesis that plankton animals remain within an optimum light intensity to make their vertical migrations (Russell, 1928). Because at such depths the light intensity is very low and very diffused, the mechanism by which fish are retained within a layer must be effectively a distribution of photokinetic movements.

4. *Some specialized observations*

There are layers outside the deep ocean which are of interest, in that conclusions drawn from their investigation are useful in the study of the Deep Scattering Layer. One of the earliest studies of this type was that of Burd and Lee (1951) who found a layer extending eastwards from Start Point in Devon, in southern England. Three hours before sunset it lay between 20 and 36 fm and 2 hr later it had risen to 5 fm and then it dispersed. At 0100 hr, the layer reformed at 12 fm, by dawn it was found at 33 fm and in daylight it remained close to the bottom. Hauls made with a Petersen Young Fish Trawl in and out of the layer showed that 70 % of the scatterers (i.e., animals found exclusively in the layer) consisted of pilchards of 11.5 mm in length and 10 % of *Crystallogobius* sp. of 12.5 mm in length. The latter had air bladders, 3 mm across and the pilchards had just acquired air bladders. The echo sounder used had an operating frequency of 14 kHz and it was quite possible that the larvae, at a density of 6 per m^3, had swim bladders which resonated to the transmitted frequency. Dragesund and Olsen (1965) have described layers of young fish, cod, redfish, herring, and long rough dabs, in the Barents Sea, which have been detected with frequencies of 25–37 kHz; it is possible that there is a frequency shift in signal as the baby fish layers rise towards the surface; perhaps the techniques of frequency analysis and estimation of volume scattering coefficients by frequency bands should be applied to the problem.

Another specialized study of value has been made in Saanich Inlet on Vancouver Island, British Columbia, Canada. At depths of about 100 m a diffuse layer was found with single scatterers and small shoals interspersed amongst it (Bary, 1966). The animals were taken with a high speed sampler towed at 5–6 knots, from which it is unlikely that euphausids, for example, could dodge. Bary found that euphausids (*Euphausia pacifica* Hansen) in the layer reached a density as high as 280 per m^3, and amphipods (*Cyphocaris challengeri* Stebbing, *Orchomonella pinguis* Boeck) 60 per m^3. So, it might have been expected that the diffuse layer, sampled at 12 kHz, might have been a layer of euphausids and amphipods. But there was no correlation between the depth of the animals and that of the diffuse layer. Hence, it is very unlikely that the oceanic layers can comprise any euphausids (or amphipods) at all in the acoustic sense, because the oceanic density of such animals is very much lower than that found in Saanich Inlet with the Bary sampler. It is true that Cushing and Richardson (1956) did find a concentration of 600 per m^3 of euphausids in the North Sea, with a Gulf III sampler which was recorded on an echo sounder. But Kinzer's (1969) estimates with the same type of net in the eastern Atlantic and Indian Ocean were very much lower, averaging 0.3 per m^3.

Another special study is that made by Lenz (1965) on the discontinuity layers in the Baltic. In this area there are differences in density of up to 5 or 10 units of σ_t. Some layers of detritus appeared to scatter sound back to the receiver when none was reflected from the density discontinuity. But the sharper discontinuities were recorded and probably reflected 0.0001 to 0.016 % of the incident intensity.

Thus, away from the ocean, studies relevant to the problem of the Deep Scattering Layer have been carried out. The sensitivity of echo sounders to physical discontinuities has been defined and, as expected, the sharpness needed to return a signal is much higher than any to be found in the deep ocean. Where euphausids are found at densities great enough to produce signals (on the basis of theoretical calculations) the layer is diffuse and its depth may not be correlated with that of the animals. Lastly, in shallow seas, the problems of the layers of larval fish are probably those of the Deep Scattering Layer in microcosm and perhaps the same physical principles of frequency analysis should be applied to their solution.

5. *The Deep Scattering Layer*

One of the hopes expressed by the early workers was that the layer would provide an enormous reserve of protein. But the fish are small and live as much as 10 m apart; the R.V. *Walther Herwig*'s catch of many thousands of fish, after towing her very large mid-water trawl for 1 hr, was poured into two buckets. The biological importance of the layer is that it is an extra trophic layer between the plankton animals and the tuna. Hence the reserve of protein, if needed, would have to be taken from the plankton, at some expense.

The scientific interest in the Deep Scattering Layer is threefold: (a) as an acoustic problem, (b) as a problem in behavior, and (c) as an ecological problem. The acoustic problems are centered on the resonance of swim bladders and the estimation of volume reverberation. and the absorption of sound by the components of the fish body. Linked to this problem is the purely physiological one of how the retia mirabilia of the swim bladder manages to pass the quantities of gas quickly enough to make the migration. The behavioral problem appears to be one of photokinesis and how the small speed differences, that might be generated in the continuous vertical movement of the isolumes, are so distributed as to retain the layer in position. Presumably it is a statistical problem, in which measured differences in speed over very short time periods would play an essential part. The ecological problem is possibly not yet formulated fully; it is really that of energy exchange between trophic levels in a vertically partitioned ocean. Riley *et al.* (1949) and Steele (1956) have formulated the flow of primary production vertically, but no such theory is available for secondary production or for the complex system of layers which constitutes the third trophic level in the ocean.

The possible scientific problems listed above are not as separated as the last paragraph might suggest. Each is linked to another because a solution in one area is also a solution in another and so advance in the study of the Deep Scattering Layer is to be expected as scientists work on it at sea.

CHAPTER 7

THE USE OF MORE COMPLEX ACOUSTIC INSTRUMENTS

BECAUSE sonars are used for purposes other than fish finding, machines for these purposes are sometimes adapted for detecting fish. Sonars are also developed for particular purposes in fisheries and specialized acoustic equipment is used on trawls. In this chapter is described some of the more advanced equipment with the results useful to the study of the abundance of fish in the sea.

1. *Development of echo sounders*

There is a great variety of echo sounders developed for particular markets. The steered narrow beam development of the Kelvin Hughes Humber gear (Pearce and Philpott, 1970) probably represents the fullest development of the traditional echo sounder for the purposes of finding fish. Here we are concerned, however, with the exploitation of those principles by which better detection has been achieved and which are useful for the estimation of fish abundance. The main development in recent years has been in the direction of high frequency equipment because of its higher resolution in angle and in range for a given size of transducer. But in few commercial machines is the full capacity for resolution thoroughly exploited. Combined with a large and fast recorder and a high pulse repetition frequency, a high degree of resolution in time is also possible. Such equipment suffers from the relatively high attenuation of sound with range, but high frequency transducers have a higher cavitation threshold and so more power can be transmitted.

The trawl transducers have been described in some detail because they might be used to describe the processes of capture. At the present time, fisheries biologists depend upon the catches of fishing vessels for their description of populations, because the catch per unit of fishing effort (or time spent fishing) is a proper index of stock. Yet it can be biased and it should be investigated directly so the biases can be eliminated. Echo-sounder observations should give the true stock density and the ratio of this to the catch per effort or apparent stock density would measure catchability, itself the ratio of fishing mortality to fishing intensity. There are three possible methods by which the catchability coefficient could be measured directly, with Dowd's method (described in Chapter 4), with the acoustic arch, or with the ARL scanner (described below). With each method, the aim is to measure what proportion of stock is taken by the trawl in a single haul.

Transducers are sensitive to propeller noise and to the bubble layer underneath the ship. Both effects can be mitigated by towing the transducer on a short cable alongside the ship. Thus the sensitivity and reliability of the whole system is increased; as noted earlier, the towed transducer lends itself more readily to convenient calibration. There is no reason why transducers should not be towed at greater depths and even quite close to the bottom for some specialized purposes. One of the difficult problems facing fishermen and fisheries biologists is how to find and catch fish on rough ground where they may be hidden. A narrow beam, short pulse length transducer, flown over rough ground, has a much greater chance of detecting fish in the hollows than has a conventional echo sounder.

Both at the Fisheries Laboratory in Lowestoft and at the Marine Laboratory in Aberdeen, high frequency equipment has been developed to exploit such characteristics to the full. As described in earlier chapters, fish tend to be recorded as individuals and so with a pulse height analyzer or a Simrad integrator it is possible to record size distributions by range gates. Perhaps in the future high powered versions of these machines will be used to sample fish populations in deep water with a much greater degree of resolution in range, angle, and time than is possible at the moment. Further, because the transducers are smaller at higher frequency, the use of arrays and stabilization procedures is less inhibiting than at the frequencies employed at the present time for fish detection.

2. *Trawl transducers*

The headline transducer, or netzsonde, is used commercially as well as on research vessels and a natural development is to use more transducers firing in different directions. Schärfe (1968) has described a multi-netzsonde for this purpose; transducers fire down, up, and forwards from the headline, down from the square and across the mouth of the trawl from the boards, bridles, and wings. So far it has been used primarily for the measurement of trawl dimensions (gape; height of headline; distance of net from bottom and surface; distances between wings, bridles, and boards). To study fish behavior comprehensively with such equipment would require a complex display equipment, with a separate recorder for each transducer. Then because of the three-dimensional nature of the received information an ideal display would be three-dimensional to be immediately intelligible.

Another device is the acoustic arch which is being developed at Lowestoft. Transducers working at 100 kHz, with wide beams are slung along the headline in such a way that their beams overlap. They are fired in sequence so that a single recorder can receive all the information. Fish may be recorded on two transducers, but the double record is so obvious that it does not matter. Hence a complete record of fish passing beneath the headline is available. Because fish may move back and forth beneath the headline it may be necessary to tilt some transducers forward and some backward so that forward movement can be distinguished from backward movement; then the area on the paper record covered by the different transducers must be indicated. So the catch in numbers can be compared with the signals recorded by the arch; there is no reason why in future, size distributions should not be compared using acoustic methods. Further, it might become possible to fly an arch between the trawl boards and so make further comparisons of the catchability of the trawl.

3. *Development of traditional sonar*

In principle the sonar used for detecting submarines during World War II has not developed. Today, the transducers can be tilted as well as trained and they can be trained automatically to a variety of different programs. Simrad have made a twin-frequency sonar for research purposes originally designed for finding the herring in the Norwegian Sea on their passage from Iceland; it is calibrated, signals can be measured, and it has a tape recorder for storing them at sea. Simrad also made a special rig for changing transducers at

FIG. 83. Survey of rock outcrops using the side-scan sonar (Chesterman *et al.*, 1958).

sea, because in the Icelandic herring fishery many have been broken off by submerged logs, which have presumably drifted from Russian rivers.

The traditional sonar had reached its limit in range and in resolution during its military development. During the fifties and sixties it was employed by Norwegian and Icelandic fishermen to very great effect in catching herring with purse-seines. But for quantitative purposes it is often limited by the refraction of its rays by thermal layering in the sea, whereas the echo sounder which transmits normally to the layers is not affected. Although the development of such a sonar has played a considerable part in fish detection for fishermen, it is not recorded here because our real concern is with the development of acoustic methods to establish the abundance of fish independently of the fishing fleets.

4. Side-scan sonar

A transducer is fired sideways from a ship, with a narrow beam in the horizontal plane, and in the vertical plane it has a very wide beam with many side lobes. This is a fan-shaped beam and as the ship steams forward the fan extending from the side examines the seabed and the mid-water from extreme range to the area directly below the ship. The first equipment (Chesterman et al., 1958) was designed to detect mines on the bottom and found an immediate use in the geological survey of the seabed because the bottom features were shown so clearly. Figure 83 shows an early survey of outcrops in Weymouth Bay; it will be seen that the picture of the rock formations can be interpreted in a very simple manner. Further, because the fan beam covers an area described by the ship's track and the maximum range abeam of the ship's course, the display is effectively a plan of the seabed. So long as the ship is navigated accurately, it becomes possible to chart areas of seabed. The thin line on the left of Fig. 83 near the transmission mark is a true indication of the depth of the sea and so the chart is calibrated in depth along one transect.

Geologists make considerable use of this form of equipment and there are beautiful underwater pictures available of harbors, wrecks and pipe lines. Although the original equipment was mounted on ships and was tiltable, the modern analog is towed and can be tilted by remote control. For any angle of tilt the maximum range may differ and the contrast in the picture may change; so for survey, a constant angle of tilt should be used, but any angle could be used for the closer examination of particular features. Recently, the geological use of the side-scan sonar has been much extended by the introduction of Gloria at the National Institute of Oceanography (Rusby et al., 1969). The transducer at 7 kHz has 144 elements, an output of up to 60 kW into the transducer, and an effective horizontal beam of about 2°; it can be towed at depths as great as 200 m and it has an effective horizontal range of about ten miles or more.

There is an obvious application of side-scan sonar to fisheries, first, to survey pelagic fish and secondly to examine the nature of the seabed. Figure 84 shows a side-scan sonar chart of sprat shoals in mid-water; not only can the total distribution be seen at a glance, but also the sizes of shoal, at least in plan, can be readily estimated. An ordinary sonar could be used in the same way, but, as the beam tends to be rather wide in the horizontal plane, the volume sampled by the beam increases with range; with the narrower beam of the side-scan sonar this effect is minimized and so range and shoal size are more properly estimated. But Smith (1970) has overcome many of the difficulties by using a middle section of the beam in ranges within which the errors are minimized and which he has corrected. Figure 85 shows differences in ground observed with a side-scan sonar; not only are gross

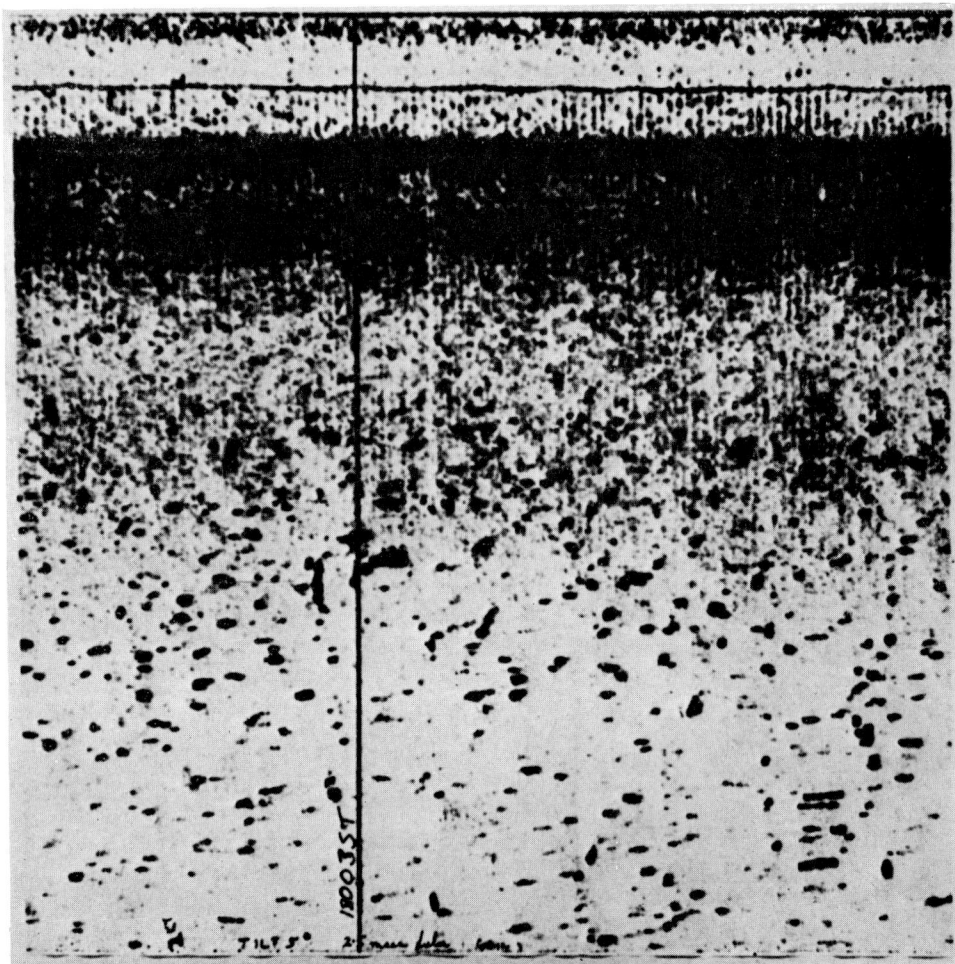

FIG. 84. Survey of sprat shoals with the side-scan sonar (Cushing, 1963).

formations like rock outcrops and pipe lines readily detected with such equipment, but so are small differences in ground due to formation or to differences in the acoustic sizes of materials. Fishermen are interested in rough ground, which can only be defined as ground which tears the trawl. The ordinary echo sounder which described a depth profile does not always describe rough ground very well. Pinnacles and raised ground are shown and rock can be distinguished from mud by the length of the trace beneath the bottom signal. But fishermen would delight in finding gullies and flat areas described for them in the midst of rough ground. Many fishermen can interpret the echo sounder record as an indicator of rough ground but they need the plan presentation given by the side-scan sonar. If the echo sounder were run at the same paper speed as the side-scan sonar, then the depth indication on the latter would be exactly correlated with the echo-sounder record. Then an extension can be made in imagination from the record of one machine to that of the other and the best use obtained from both machines. With practice, a side-scan sonar could be used to considerable advantage. Fisheries biologists would also find the instrument of great use in

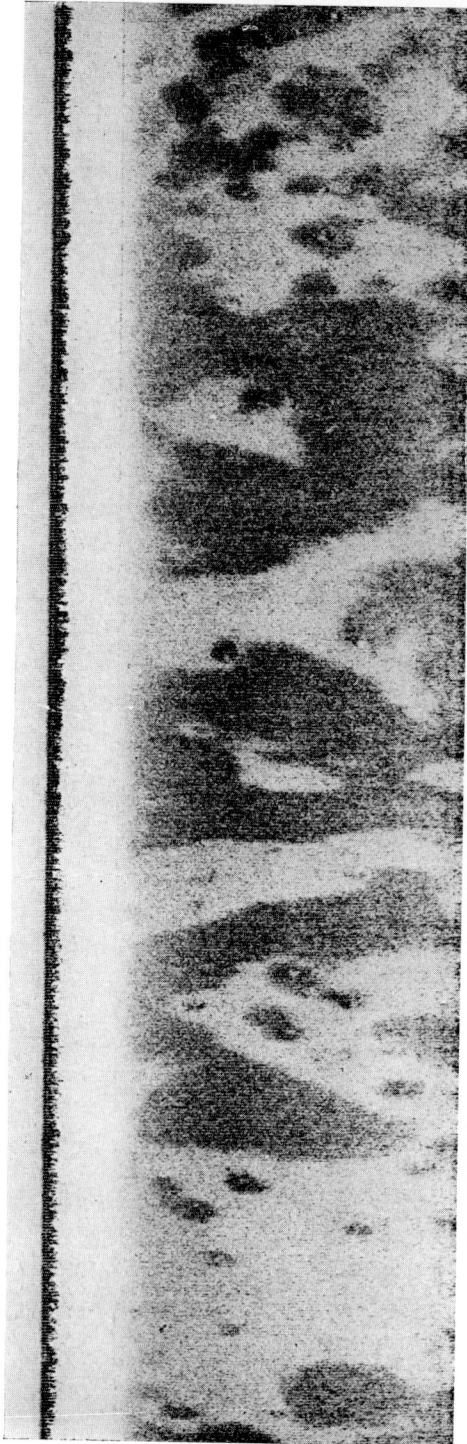

FIG. 85. Differences in ground observed with a side-scan sonar (with kind permission of Kelvin Hughes Ltd.).

charting the distribution of different substrates, which might be expected to support different foods and different food fish. It is one of the mysteries of present day fisheries research that the side-scan sonar has not been exploited to its full use by fishermen and fisheries biologists.

5. *Continuous transmission frequency-modulated sonar (CTFM)*

This equipment, which is made by Straza Industries, San Diego, California, U.S.A., has been developed from a mine hunting equipment; there are two forms used in fisheries work and some smaller versions are working on various submersibles. Between an upper and a lower frequency sound is transmitted continuously, from the upper to the lower, and then, immediately, the sweep starts again and the process continues in a sawtooth manner. In echo, sound returns after an interval of time governed by the range of the target, so the difference frequency between transmitted and received frequencies indicates range. Frequencies are analyzed by a bank of 100 filters. There are two transducers, a long range one working out to 1600 m and a short range one to 500 m. The frequency band for the first lies between 52 and 32 kHz and that for the second between 290 and 260 kHz. The long range transducer has an effective horizontal beam of 6° and a vertical beam of 15°; the short range one has a two-dimensional array of elements from which horizontal, vertical and conical beams can be selected. The conical beam is 0.6 × 0.6°, the horizontal 0.6°, with 15° in the vertical plane and the vertical beam is also 0.6° with 15° in the horizontal plane. The whole equipment can be trained automatically and tilted to 90°. This equipment has the highest resolution in angle that is obtainable, but because the transducer is not stabilized against pitch, roll, and yaw, the high resolution cannot be fully exploited. It has a joystick control which provides a very precise means of handling the transducer as the information comes in, but when one switches from long range to short range, the long range movements are transferred to short range action and the target is readily lost.

There are two advantages in the system. The first is that the horizontal scan rate can be as high as $30°$ $\sec^{-1}$, which is as fast as a radar scan. Range resolution is a function of the number of filters employed and increases with range; it is 1 m at 100 m range and 16 m at 1600 m range; at long range the fine resolution may not be needed but the swift flow of information in bearing because of the high horizontal scan rate may be very useful. Indeed at such ranges, the only information needed is precise bearing; further, information on targets at bearings other than that under examination is much less valuable to a fisherman than to a submariner, because he hasn't got to protect himself against attack. The other advantage is that information on signal strength on a given target is collected very quickly, as function of the continuous transmission and it is displayed on a storage oscilloscope and averaged. During the time that a signal travels to a target and returns with a conventional sonar, many signals have been received with CTFM equipment; if one needs much signal information for averaging procedures as suggested in Chapter 4 then this equipment has considerable advantages over conventional sonars. If one compares whole systems it can be shown that the information flow is the same whether CTFM equipment or a sector scanner (see below) is used. But for selected pieces of information there may be advantages in using one form rather than another.

When a sonar is trained from fore and aft to the beam and back again, the frequency shifts slightly because of the differences in the distances in transmission and reception due to the ship's speed; the difference frequency is a Doppler frequency. Because it is frequency modulated CTFM equipment has a device called an "own Doppler nullifier", whose function is self explanatory. Hester (1967) has used the same equipment without the nullifier but with a frequency analyzer which records the frequency shifts. Three classes of target were found; stationary ones, narrow band Doppler (directional movement), and broad band Doppler (complex movement relative to the transducer), and the latter were found to be characteristic of fish shoals. The details of equipment specifications differ a little from those described above, with a different source frequency (75–60 kHz, but 70 kHz in the Doppler mode) and receiver beam angles of 4° horizontally by 10° or 20° vertically.

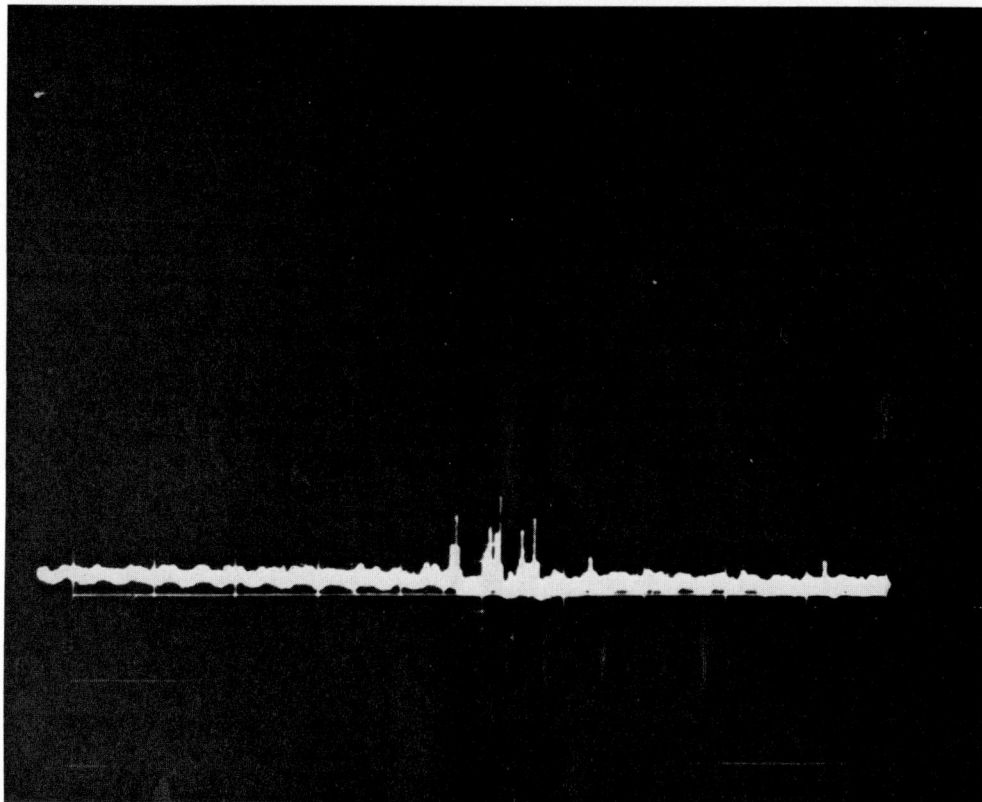

Fig. 86. Doppler signals from a school of "bait fish" at a range of 50 yd (Hester, 1967).

The Doppler spectrum was displayed on a CRT and the frequency resolution after using a Deltic analyzer was as high as 1.67 Hz in ±300 Hz. Figure 86 shows Doppler signals from a school of "bait fish" at a range of 50 yd; frequency is measured left or right of the center as up (left) or down (right) Doppler as indicating movement towards or away from the transducer. A single frequency is recorded as a single spike, whereas multiple frequencies were displayed as many spikes of lower amplitude. The equipment is sensitive to differences due to ocean currents, sea state, pitch and roll, and vessel velocity, most of which can be eliminated with a proper combination of luck and judgement. Figure 87 shows a single fish

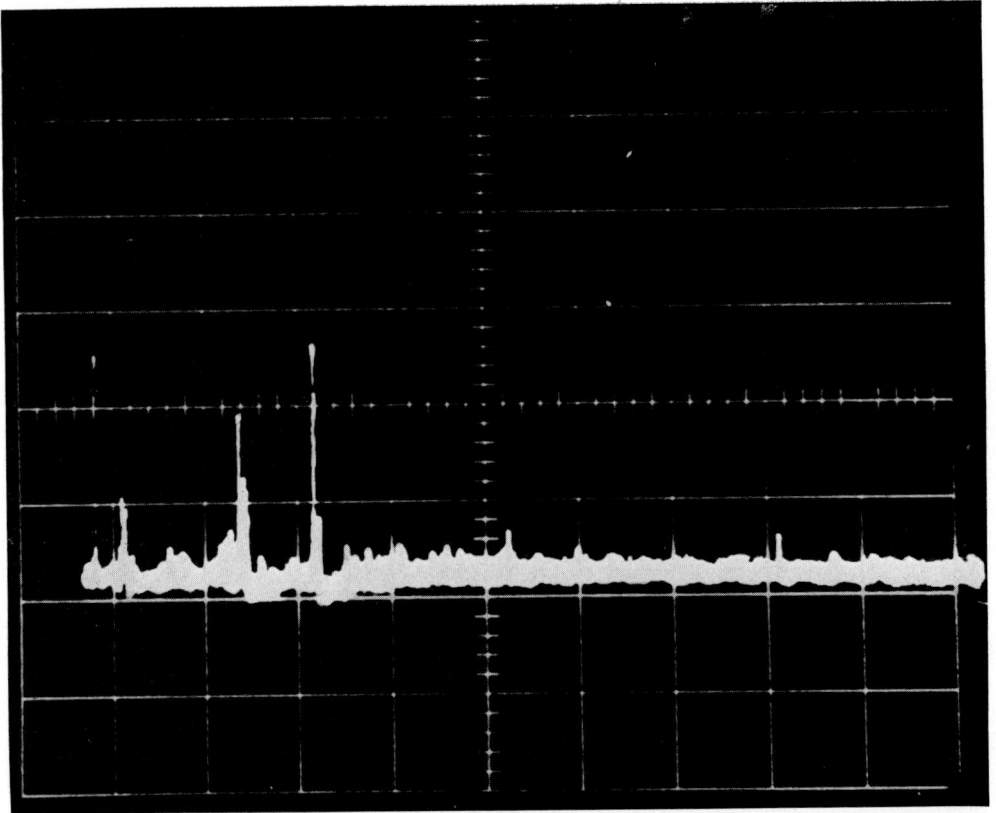

Fig. 87. Doppler signals from a single fish approaching the ship at an apparent speed of 115 cm sec^{-1}
(Hester, 1967).

exhibiting up Doppler, i.e., approaching the ship at an apparent speed of about 115 cm sec^{-1} (if the maximum apparent velocity is 321 cm sec^{-1} on the ± 300 Hz scale).

The CTFM equipment has three advantages which are unexploited by other means: the high rate of training, the rapid averaging, and the use of Doppler. It is likely that these advantages will be exploited in the future by forms of frequency modulated equipment.

6. Sector scanners

Another solution to the problem of searching the sea bed is the sector scanner or within-pulse scanner. A wide beam is transmitted and echoes are received by a transducer array with many channels. The spherical wave from the target is received at progressive intervals of time on different channels. The differences in time can be resolved electronically to give the bearing of the target. The acuity of the system is determined by the beam width of the whole array. The narrow beam is steered electronically, the wide angle insonified in transmission being scanned in angle in reception. The electronic resolution of the time differences at each sector can be carried out in a number of ways. Tucker and Welsby (1960) converted the phase difference at each channel into frequencies and then resolved the bearing of targets using delay lines. Voglis and Cook (1966) modulated the difference in frequency directly using transformers. Both forms of *sector scanner* present their signals on a rectangular screen, bearing on range. Because the true plot is in polar coordinates, distortion of shape appears in the picture.

The first biological use of the sector scanner was that of Harden Jones and McCartney (1962). They used Tucker's first prototype from R.R.S. Discovery II. Fish shoals were tracked past the anchored ship, the sector scanner transmitting on the beam. Contemporaneous current measurements were made and most herring shoals were shown to be drifting with the tide. Figure 88 shows a polar distribution of vectors of shoal movements observed from the anchored ship. Most of the shoals were shown to move in about the same direction as the tide, but a minority moved against the tide. The same result appears from the results of a study of young herring being swept by the tide past the inlet of a power station in the Firth of Forth (Welsby et al., 1963), using the University of Birmingham equipment. The same equipment has been used to study changes in fish behavior in a tank by night and day (Welsby et al., 1964); an interesting method of processing the films was introduced by superimposing frames and the trace-to-trace correlation accented the movement of fish shoals as distinct from bottom signals and noise.

The ARL scanner has the following specifications:

Operating frequency	300 kHz
Pulse repetition rate	4 per sec
Transmitter characteristics:	
Array	curved
Horizontal beam width	30°
Vertical beam width	5°
Nominal range resolution	8 cm
Receiver characteristics:	
Array form	linear
Number of elements	75
Horizontal beam width	0.33°
Vertical beam width	10°
Scanning rate scans sec^{-1}	10,000
Scanning sector	30°
Number of independent channels	75

As reported by Cushing and Harden Jones (1967), the effective range was about 200 yd. The display was photographed on a 200 yd range or an expanded 50 yd range and the films were examined either as cine-film or as single frames. From echo-sounder records, shoals had been regarded as spheres, ellipsoids, or lumps. Figure 89 shows some examples of spheres, loops, crescents, and strings.

An interesting point is that individual fish can be recognized in the daytime shoals of pilchards. One shoal was observed for about 120 frames. It was possible to count the individual fish within the shoal. The counting error was low, despite the fact that the sampling volume of each sector range annulus is rather large, because of the vertical beam. Hence the fish in the shoal examined must have been fairly distant from each other, 3–5 fish lengths rather than one fish length. The chance of fish being detected within a sampling volume follows a Poisson distribution and because the volume decreases with decreasing range, so the chance of detection of a single fish increases (Gulland, in an appendix to Cushing and Harden Jones (1967)). Hence counts at different ranges on the same shoal can be used to determine the true number of fish in a shoal, or the packing density of the fish. The whole

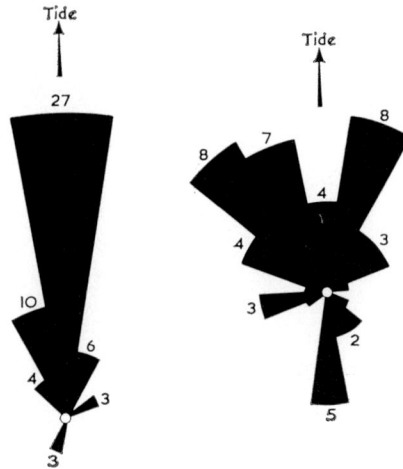

FIG. 88. Distribution of shoal vectors with respect to tidal streams (Harden Jones and McCartney, 1962).

shoal cannot always be resolved into individuals, but from Fig. 89 it can be seen that on some occasions much of the shoal is resolved; preliminary estimates of packing density could be readily made from such information. One remarkable sequence on film shows two string shoals each with outriders. When the outriders came within visual range of each other the two shoals turned and merged at about 22 yd. Figure 90 shows the complete sequence at a range of about 100 m; the two shoals were 55 m apart to start with and the whole event lasted 13 min. Close examination of the leading point of the shoals shows a scattering of individuals, as if they were searching outriders.

The scanner was also used to examine an Engel mid-water trawl towed by a research vessel. Figure 91 shows a complete underwater picture of the trawl which was shown completely on one frame. The trawl is shown in outline by echoes from its selvedges; from the otter boards, from the warps and bridles, and of course from the headline floats. Fish echoes were seen to station themselves along the warps and bridles for a number of frames. The most remarkable sequence showed fish enter the trawl at one wing, fall back to the square and cross over to the other wing and out of the trawl (Fig. 93b). When the trawl rig was analyzed on true coordinates, it was found to be wrongly rigged. Further, a single fish was seen to station itself between the otter boards of a mid-water trawl for 80 frames.

Recently, the ARL scanner has been fitted to the R.V. Clione (of the Fisheries Laboratory, Lowestoft, U.K.). The transducer is suspended from a package which rides in a tube through deck and keel; it is stabilized in roll, pitch, and yaw and it can be worked in the vertical mode as well as the horizontal one. Figure 92 shows a fish shoal in the vertical mode; it is effectively a vertical section of the sea, analogous to an echo-sounder record. The seabed is shown as a curved line on the right of the picture and the vertical section of the shoal is shown in the center, resolved into individual fish. One of the remarkable results from the last year's work is that the bottom trawl can be unexpectedly described. Figure 93 (Margetts, personal communication), shows a picture taken with the ARL scanner of a Granton trawl; headline, ground rope, dan lenos, cod end, bridles, and doors are visible;

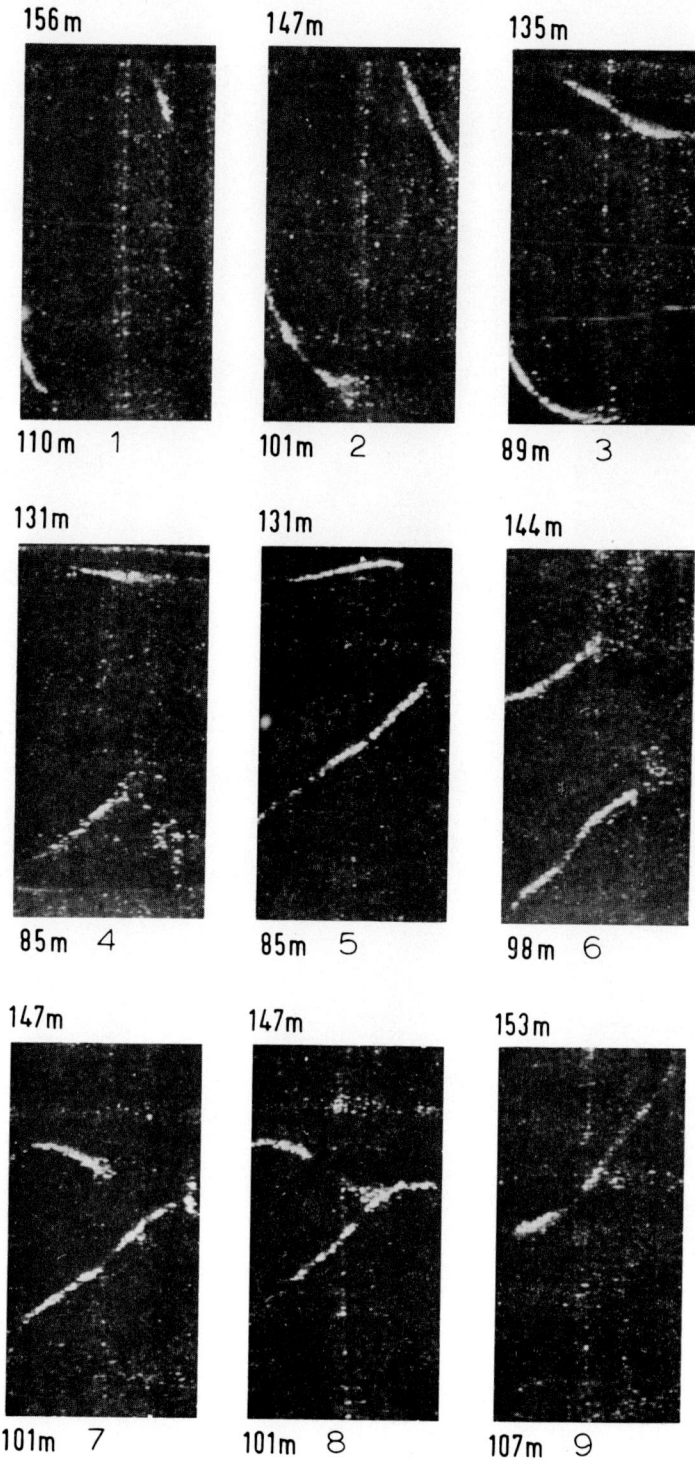

FIG. 89. Two string-like shoals approach each other and merge in 13 min at a range of 22 yd (Cushing and Harden Jones, 1967).

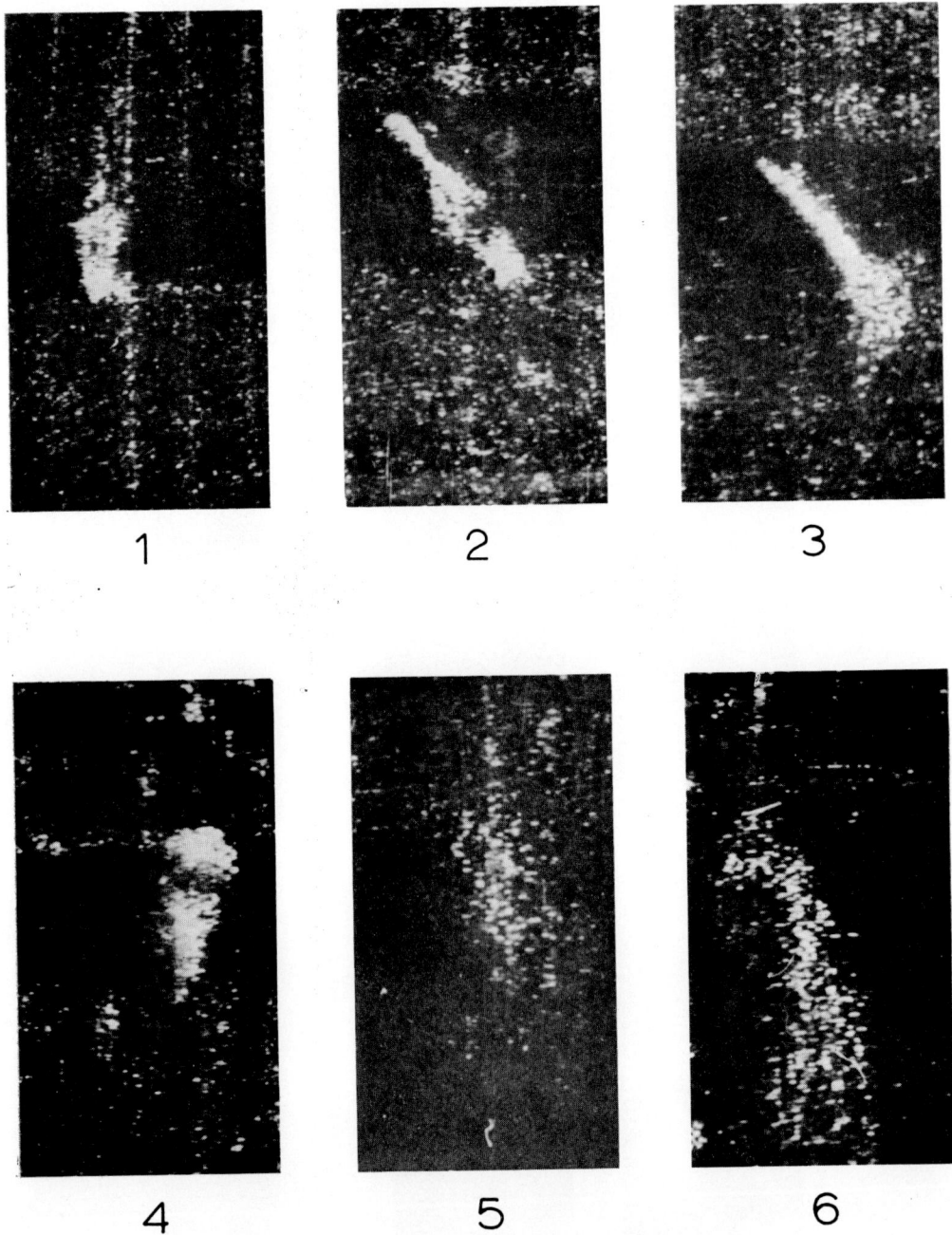

FIG. 90. A number of fish shoals recorded by the ARL scanner (Cushing and Harden Jones, 1967).

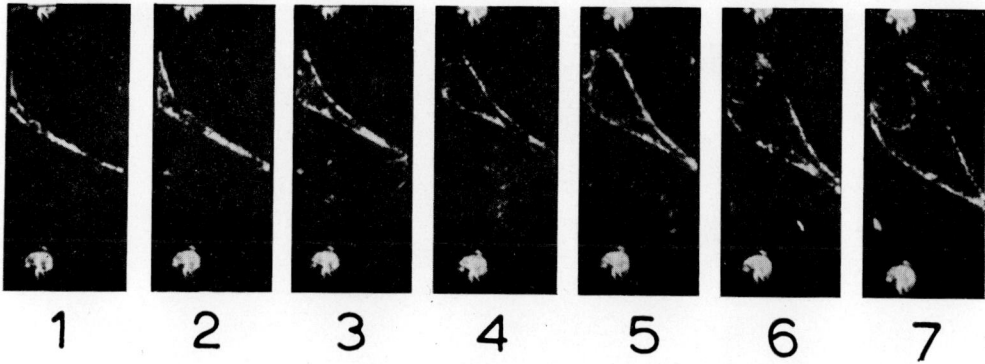

FIG. 91. An Engel mid-water trawl shown completely by the ARL scanner (Cushing and Harden Jones, 1967).

such a full picture cannot be constructed by any other means. Thus measurements of spread between the doors, of bridle angle, of headline catenary, and trawl shape can be obviously and readily made. Less obviously, measurements of headline height can be made in the vertical mode and of the door height and attack angle from its acoustic shadow. Perhaps the most unexpected points in Fig. 93 are the eddy trails streaming away from the trawl

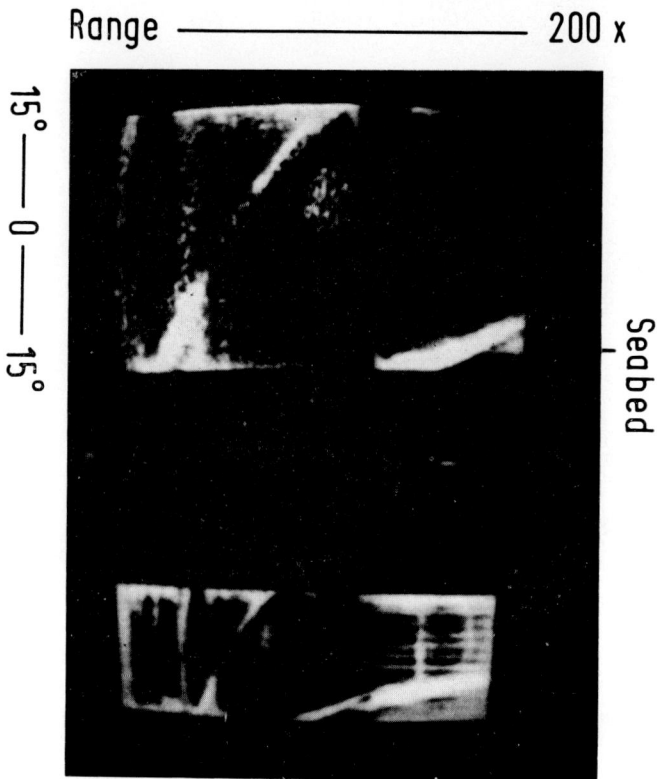

FIG. 92. A fish shoal shown in the vertical mode with the ARL scanner.

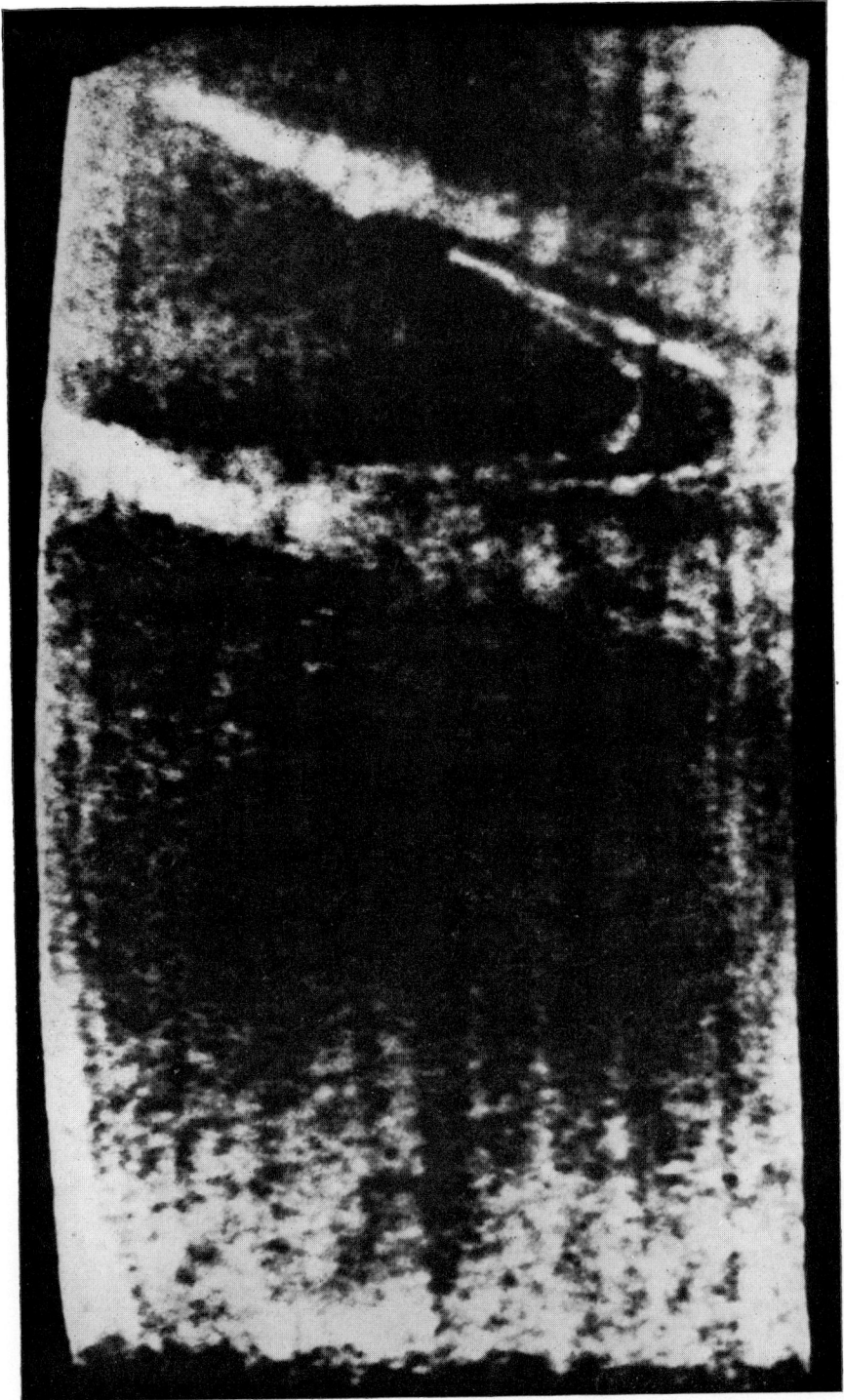

FIG. 93a. The Granton trawl as displayed on the ARL scanner (Margetts, personal communication).

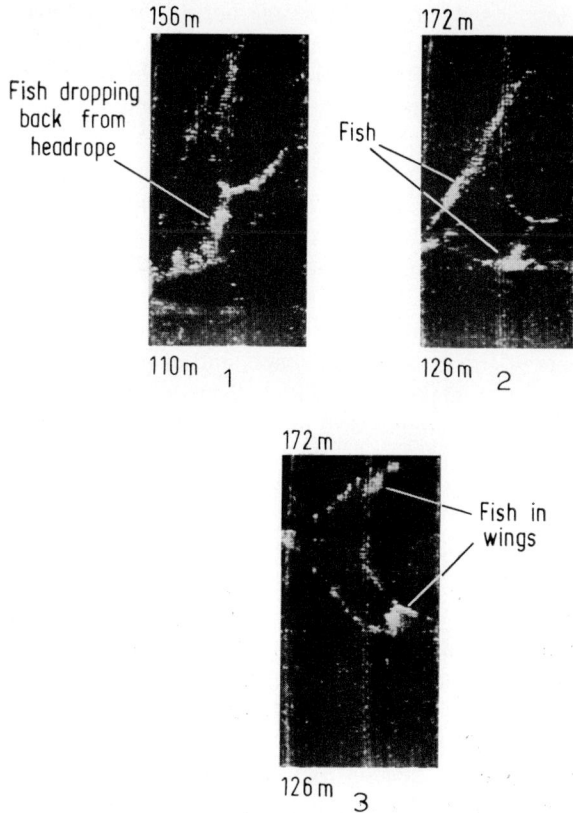

FIG. 93b. An Engel trawl with fish observed inside it.

doors. Their existence has been postulated and they have been photographed at short range, but in Fig. 93a they are fully described in height, width, length, and density. It is possible that such eddy trails play a considerable part in the responses of demersal fish to the advance of a bottom trawl; in Fig. 93b fish are shown within a mid-water trawl.

The thin bands at all ranges are bands of noise detected at particular bearings. They are generated by parts of the trawl scraping on the bottom; the doors and the ground rope, with their heavy iron bobbins, must make a lot of low frequency noise, but the dan lenos also do so. This phenomenon, if unexplained in terms of frequency, can be used to show where the ground rope touches the bottom and if one door is digging in more than the other. A somewhat related point is that noise bands are generated by the tide as it flows over sand ridges (Mitson and Cook, 1970); this is of more interest to geologists but it is possible that fish can hear the low frequency components of such noise and use it for finding their way round sand banks.

The capacity of the ARL scanner for various purposes has been enhanced recently by the invention of a transponding acoustic tag, operating at 300 kHz (Greer Walker, et al. 1970). As distinct from earlier acoustic tags it is very small (about 4.0 cm in length), nearly comparable in size, weight, and drag to the tags used by fisheries biologists for other purposes. The tag has been used on a plaice and for 15 hr the fish was followed continuously. The fish rose off the bottom and drifted with the tide when it started to flow more quickly;

later as the tide slackened, the fish settled on the bottom with a flurry of sand. There are many uses to which this technique can be put, and of these the most important for fisheries biologists are the detailed study of the migration of individuals and that of the work done by fish as they swim for long periods. The catchability of the Granton trawl is now being studied directly with the acoustic tag.

It is likely that the sector scanner is the most important acoustic equipment to come into the hands of fisheries biologists since the invention of the recording echo sounder by Wood et al. (1935). As Harden Jones and McCartney (1962) have already shown, the scanner can measure the migration of shoals and individuals in detail with respect to current measurements made at the same time. In view of the present interest in fish migration (Harden Jones, 1968), such observations are crucial. A second profitable field of research is the accurate measurements of shoal size, shoal shape, and of packing density. Moreover, used vertically the sector scanner might be the most effective fish counter yet devised, because all the ambiguities in directivity referred to in Chapter 4 have been overcome. Lastly, there is some evidence that the responses of fish to the trawl can be studied in some detail.

7. Low frequency equipment

From the study of the Deep Scattering Layer, it is obvious that fish can be detected at considerable range and at low density if their swim bladders resonate to the transmitted frequency. But the fish in the layer are small and the question arises as to which frequency should be used to locate the larger commercial fish at considerable ranges. Unfortunately the frequencies needed are low, of the order of 1–2 kHz, and in this region the transducers are very sensitive to noise of all sorts. There is a geological instrument which operates at these frequencies, the "boomer" or seismic profiler. It transmits at about 1 kHz and is used for the survey of deep sediments (Hersey et al., 1961). With this equipment signals have been recorded from fish in mid-water, probably pilchards (McCartney et al., 1965); the swim bladders of pilchards would resonate between 0.6–1.9 kHz. However, there are difficulties both in the transmission of power at a precise frequency and in the interpretation of the received signals, particularly from shoals of fish. In Chapter 4 the use of analogous equipment was referred to in the account on estimating abundance over rather long ranges. In principle such equipment could be used more extensively in fisheries research, but at the moment it is perhaps too expensive.

Weston and Revie (1971) have investigated fish echoes on a long-range sonar at low frequency (about 1 kHz), with a 15° projector and a 4° receiver. Within a narrow band the signal was frequency modulated and the received signals were examined with correlation techniques. On a display of range on time, fish echoes were detected to a range of 18 nautical miles. The traces appeared as tracks in time, which were probably due to fish shoals; an average density of 2 shoals km^{-1} was recorded. Most tracks were observed in summer and all dispersed at night and this diurnal variation was correlated with the times of sunset and sunrise. Perhaps the most remarkable result was the variations in range with time which were closely correlated with tidal periods for days at a time. Hence the shoals of pelagic fish drift with the tide for such periods of time, as expected by Harden Jones (1968). Figure 94 shows traces of fish at ranges up to 16 miles for 30 hr and shows their drift with the tide.

The equipment used is that briefly noted in Chapter 4 where it was noted that estimates of fish density could be made from variations in sound propagation. From the echo results,

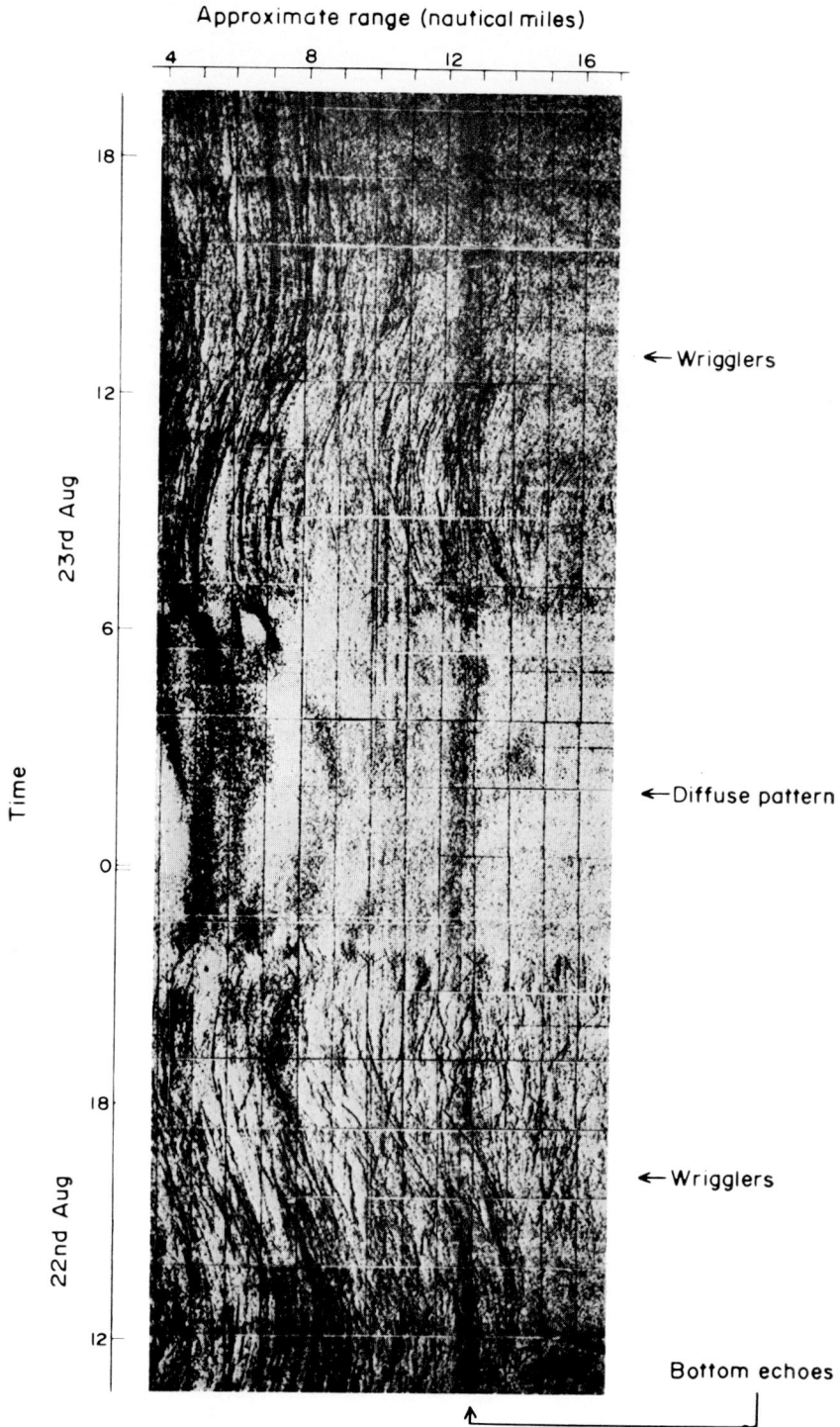

FIG. 94a. Fish traces observed at long range with a low frequency sonar, showing the tidal movement of shoals.

Approximate range (nautical miles)

FIG. 94b. Fish traces observed at long range with a low frequency sonar. Detailed observations at a range of about ten nautical miles.

it is likely that the swim bladders were resonating to the transmitted sound. From the received signals, Weston and Revie conclude that the resonant swim bladders were in fish of 23 cm in length, possibly pilchards. The average fish shoal had a target strength of $+5$ dB. If the fish were packed at a density of one length apart, 2 shoals km^{-2} amounts to a density of 6.6×10^4 km^{-2}; by sound propagation, the authors deduced a density of 4–6×10^4 km^{-2}.

Such equipment, although very expensive, has considerable application in fisheries biology. The behavior of fish in response to tidal streams can be investigated on a very large scale. The abundance of pelagic fish can be examined on the same scale, for considerable periods. The point is that this is the scale on which the basic statistics of fisheries biology are taken from the sea, in catches and stock density.

8. *The use of echo sounders and the more complex instruments in fisheries biology*

The various forms of acoustic instrument were developed originally for navigational or military purposes. The echo sounder for fishing purposes is today very different in character from the navigational instrument, in power output, in transducer design, and in display. The evolution has taken place in response to the needs of fishermen and fisheries biologists. The later and more complex instruments have not yet diverged much from the design needed for their original application, but in the future they must do so. Whatever the developments in other directions, it is likely that acoustic methods of search, detection, and measurement will remain the staple ones in the sea, if only because sound travels far in the sea whereas light is scattered. It is true that pulsed lasers may prove of great benefit in the future, but for most purposes it is likely that we will continue to depend upon acoustic methods.

For pelagic fishermen the conventional sonar is of great value in searching because the search field is increased by many times. At the present time it is adapted to purse-seine fishing, with a beam that broadens at short range to embrace a large shoal if needed. It is possible that sonars with much higher scan rates will replace the conventional machines, but the purpose will remain the same, to find fish shoals in the broadest area. Under homothermal conditions there is no reason why the methods described in Chapter 4 should not be developed for their specialized sonar use where the fishermen are in advance of the fisheries biologists. However, in general terms, it is likely that sonar will be used to indicate the presence of fish rather than always make an estimate of its quantity.

The problem of estimating the proportion of stock caught by the trawl (or any other gear) is a very important one. The simplest way of regarding a trawl is to think of it as sweeping an area, catching randomly distributed plaice as if they were inanimate objects. Then catch is proportional to the area swept, but no fisheries biologist dare think in this simple way because fish live in patches and they dodge the trawls. The only way in which the problem can be solved at the present time is to correlate differences in death rate with differences in fishing intensity, which requires large quantities of information collected ponderously from year to year. An alternative way is to measure the catchability coefficient at sea with different trawls and different species, under different conditions. Netzsondes and their successors, acoustic arches and scanners (with acoustic tags) will probably be used for this purpose, and the components of variance of catchability can be analyzed directly. The quality of information in fisheries biology needs radical improvement and this is one of the ways in which it could be done.

Fish often like to live on rough ground and although they can be caught there with lines or bottom gill nets (as long as it is not too rough), the rough ground remains a form of sanctuary. Fishermen would like to chart the roughs in order to work as closely to them as they possibly can. Fisheries biologists would like to work in them to understand why fish gather there in the first place, whether it is a search for food, a more varied seascape, or a shelter from quirks in the current structure. Side-scan sonars could be used for these purposes. At the present time they are rather cumbrous and expensive, but they need not be so in the shallower waters where they would be most valuable.

The sector scanners reveal a view underwater that films and television cannot reach, but the apparent array of detail is deceptive. However beautiful the picture may be, it is noisy and many of the desired details are lost in the noise. Fishermen may not use such equipment just yet, but fisheries biologists will use it for a variety of purposes. Of these perhaps the most spectacular is the description of gear in its underwater form which cannot be achieved in any other way. Possibly, the response of fish to various gears will be described, particularly with the use of the acoustic tag; the actual behavior mechanism will not be described, but the movements of the fish in response to the gear should be well portrayed. But some forms of fish behavior can be described: the responses of migrating fish to water currents, their movements in response to changes in their visual fields, and so on. The shapes of fish shoals and their sizes will be fully described and possibly estimates of packing densities will be made. Such work will be of considerable value in the acoustic estimation of fish abundance.

Echo sounders have been used for thirty years or more to indicate the presence of fish and to chart their relative abundances. More recently the possibility has appeared of measuring absolute abundance by acoustic methods and the instrumentation appended to the echo sounder will become more complex; voltage integrators and pulse height analyzers are already in use. The object of such work must be to obtain size distributions of fish per unit volume in the sea within specified range intervals. Provided that the species composition is known, by capture, an independent method is then given of estimating the stock at sea. Fisheries biologists can estimate stock independently of catch statistics by counting the eggs laid in the sea, but such methods are limited to spawning grounds. In exploration it would be desirable to estimate stock before exploitation starts on grounds other than spawning grounds. Some of the disasters in over-exploitation of herring stocks in the north Atlantic and of stocks of other species in the north Pacific could have been avoided if methods had been available of estimating stock before exploitation. The same point applies to the exploited stocks under some circumstances; if scientific failure occurs, an independent method of estimation is needed. There is, however, no need to replace the methods based on fisheries catch statistics; such methods can be improved by increasing the quality of information, but the quantity obtained from the commercial fleet must always be vastly greater than that obtained by a single research vessel.

BIBLIOGRAPHY

AASEN, O. (1962) "On the correlation between the arrival and spawning of the Norwegian winter herring" *J. Cons. Intern. Perm. Explor. Mer.* **27.2**, 162–166.

AHLSTROM, E. H. (1966) "Distribution and abundance of sardine and anchovy larvae in the California current region off California and Baja California 1951–64: A summary" U.S. Fish and Wildlife Serv. *Spec. Sci. Rep.* **534**, 71pp.

ALVERSON, D. L. (1967) "Distribution and behaviour of Pacific hake as related to design of fishing strategy and harvest rationale" FAO Conf. Fish Behaviour, *FAO Fish Rep.* **62.2**, 361–376.

ALVERSON, D. L., PRUTER, A. T. and RONHOLT, L. L. (1964) "A study of demersal fishes and fisheries of the northeastern Pacific Ocean" H. R. MacMillan Lectures in Fisheries, *Inst. Fish Univ. Brit. Columb.* 190 pp.

ANCELLIN, J. and NÉDELÈC, C. (1959) "Marquage de harengs en Mer du Nord et en Manche orientale (Campagne du Président Théodore Tissier) Novembre 1957" *Rev. Trav. Inst. Pêches. Marit.* **23**, 177–200.

ANDERSON, V. C. (1950) "Sound scattering from a fluid sphere" *J. Acoust. Soc. Amer.* **22**, 426–431.

ANDREEVA, I. B. (1965) "Acoustical characteristics of sonic scattering layers in ocean" Proc. 5th Intern. Congr. Acoustics Paper E.68.

ANDREEVA, I. B. and CHINDONOVA, YU, G. (1964) "On the nature of sound scattering layers" *Okeanologiya I.* 112–124.

ANDREEVA, I. B. and ZHITKOVSKII, J. J. (1968) "New data of the deep sea scattering layers" *Okeanologiya* **8.5**, 931–932.

ANON. (1965) "Report of third meeting of the Atlanto-Scandian Herring Working Group" *Cons. Intern. Explor. Mer. Mimeo* **19**.

ANON. (1969) "Calculation of sampling cross distance of the beam by counting the numbers of echoes in each fish trace" Tech. Rep. ICES/FAO Acoustic Training Centre, *FAO Fish Rep.* **78**, 27–28.

AOYAMA, T. (1960) "Echo patches of hair tails observed at the middle area of the E. China Sea, in February 1959" *Bull. Jap. Soc. Sci. Fish.* **26**(12), 1162–1166.

AYUSHIN, B. N. (1963) "Abundance dynamics of herring populations in the seas of the Far East and reasons for the introduction of fishery regulations" *Rapp. Procèss. Verb. Cons. Intern. Explor. Mer.* **154**, 262–270.

BAINBRIDGE, R. (1960) "Speed and stamina in three fish" *J. Exp. Biol.* **37**, 129–153.

BALLS, R. (1946) "Fish on the Spotline" Marconi Int. Mar. Comm. Co. Ltd. London, 37 pp.

BALLS, R. (1951) "Environmental changes in herring behaviour: a theory of light avoidance as suggested by echo sounding observations in the North Sea" *J. Cons. Perm. Int. Explor. Mer.* **17**, 274–298.

BALLS, R. ("Peko") (1954) "Silver veins and lucky strikes" *World Fishing* **3**, 94–98.

BARHAM, E. G. (1963) "Siphonophores and the Deep Scattering Layer" *Science 140*, **3568**, 826–828.

BARHAM, E. G. (1966) "Deep Scattering Layer migration and composition: observations from a diving saucer" *Science 151*, **3716**, 1399–1403.

BARY, B.McB. (1966) "Back scattering at 12 kc/s in relation to biomass and numbers of zooplanktonic organisms in Saanich Inlet, British Columbia" *Deep Sea Res.* **13.4**, 655–666.

BEAMISH, F. W. H. (1966) "Vertical migration by demersal fish in the north west Atlantic" *J. Fish. Res. Bd. Can.* **23.1**, 109–139.

BEKLEMISHEV, K. V. (1964) "Echo sounding records of macroplankton concentrations and their distribution in the Pacific Ocean". *Trudy. Inst. Okeanologiya* **65**, 197–229.

BELLOC, G. (1935) "La co-opération de la navigation aérienne" Mem. Office, *Sci. Tech. Pêche marit.* **9**, 130–133.

BEVERTON, R. J. H. and LEE, A. J. (1965) "The influence of hydrographic and other factors on the distribution of cod on the Spitzbergen shelf" Spec. Publ. 6 ICNAF, ICNAF Environmental Symp. 225–247.

BJERKNES, J. (1961) "El Niño; a study based on analysis of ocean surface temperatures 1935–1957" *Bull Inter. Am. Trop. Tuna Comm.* **5**(3), 219–303.

BLACKBURN, M. (1956) "Sonic scattering layers of heteropods" *Nature 177*, **4504**, 374–375.

BLAXTER, J. H. S. and CURRIE, R. I. (1967) "The effect of artificial light on acoustic scattering layers in the ocean" *Symp. Zool. Soc. Lond.* **19**, 1–14.

BLAXTER, J. H. S. and PARRISH, B. B. (1958) "The effect of artificial lights on fish and other marine organisms at sea" *Mar. Res. Scot. 1958* **(2)** 24 pp.

181

BODEN, B. P. and KAMPA, E. M. (1957) "Lumière, bioluminescence de la couche diffusante profonde en Méditerranée occidentale" *Vie et Milieu.* IX, 1–11.

BODEN, B. P. and KAMPA, E. M. (1967) "The influence of natural light on the vertical migration of an animal community in the sea" *Symp. Zool. Soc. Lond.* **19**, 15–26.

BODEN, B. P., KAMPA, E. M. and ABBOTT, B. C. (1960) "Photoreception of a planktonic crustacean in relation to light penetration in the sea" *Progress in Photobiology.* Proc. 3rd. Int. Congr. Photobiol. 1960. 189–196.

BOLSTER, G. C. (1955) "English tagging experiments" *Rapp. Procès. Verb. Cons. Intern. Explor. Mer.* **140**, 11–14.

BOLSTER, G. C. (1958) "On the shape of herring shoals" *J. Cons. Perm. Explor. Mer* **23**(2), 228–234.

BOLSTER, G. C. and BRIDGER, J. P. (1957) "Nature of the spawning area of herring" *Nature, Lond.* **179**, 638.

BOSTRØM, O. (1955) "Peder Ronnestad' Ekkolodding-og- meldetjeneste ar Skreiforekomstene i Lofoten i tiden 1st March–2nd Apr. 1955" Praktiske fiskeforsok 1954 og 1955. *Arsberet. vedkomm. Norges. Fisk.* **9**, 66–70.

BOWLEY, C., GREAVES, J. R. and SPIEGEL, S. L. (1969) "Sunglint patterns: unusual dark patches" *Science, N.Y.* **165**, 1360–1362.

BRANDHORST, W. (1958) "Thermocline topography, zooplankton standing crop and mechanisms of fertilization" *J. Cons. Intern. Perm. Explor. Mer.* **24**, 16–31.

BRAWN, V. (1960) "Seasonal and diurnal vertical distribution of herring (*Clupea harengus* L.) in Passamaquoddy Bay. NB" *J. Fish. Res. Bd. Can.* **17.5**, 669–671.

BRAWN, V. (1961) "Reproductive behaviour of the cod (*Gadus callarias* L.)" *Behaviour* **18.3**, 177–198.

BREDER, C. M. (1951) "Studies on the structure of the fish school" *Bull. Am. Mus. Nat. Hist.* **98**, 1–27.

BRESLAU, L. R. (1967) "The normally incident reflectivity of the sea floor at 12 kc/s and its correlation with physical and geological properties of naturally occurring sediments" Wood's Hole *Tech. Rep.* 67–16 unpublished MS.

BROCK, V. E. (1959) "The tuna resource in relation to oceanographic features" *Circ. U.S. Fish Wildlife Serv.* **65**, 1–11.

BROCK, V. E. and RIFFENBURGH, R. H. (1960) "Fish schooling; a possible factor in reducing predation" *J. Cons. Int. Perm. Explor. Mer.* **25**, 307–317.

BROWN GOODE, G., COLLINS, J. W., EARLL, R. E. and CLARK, A. M. (1895) "Materials for a history of the mackerel fishery" *Rep. Commissioner Fish Fisheries* **1884**, 93–531.

BULL, H. O. (1936) "Studies on conditioned responses in fishes. VII Temperature perception in teleosts' *J. Mar. Biol. Assn. UK. N.S.* **21**, 1–27.

BULLEN, G. E. (1910–13) "Some notes on the feeding habits of mackerel and certain clupeids in the English Channel" *J. Mar. Biol. Assn. UK. N.S.* **11**, 394–403.

BURD, A. C. (1964) "Abundance estimates in the East Anglian herring fishery" *Rapp. Procès. Verb. Cons. Int. Explor. Mer.* **155**, 74–80.

BURD, A. C. and CUSHING, D. H. (1962) "I. Growth and recruitment in the herring of the southern North Sea. II. Recruitment to the North Sea herring stocks" *Fish. Invest. Lond.* II **23(5)**, 71 pp.

BURD, A. C. and LEE, A. J. (1951) "The sonic scattering layer in the sea" *Nature* **167**, 624–626.

CALKINS, T. P. and CHATWIN, B. M. (1967) "Geographical distribution of yellowfin tuna and skipjack catches in the Eastern tropical ocean by quarters of the year, 1963–1966" *Inter-Amer. Trop. Tuna Comm.* **12.6**, 433–508.

CARPENTER, B. R. (1967) "A digital echo counting system for use in fisheries research" *Radio and Electr. Eng.* **33.5**, 289–294.

CHAPMAN, R. P., BLUY, O. Z. and ALDINGTON, R. H. (1971) "Geographic variations in the acoustic characteristics of Deep Scattering Layers" *Proc. Int. Symp. Biol. Sound Scatt. Ocean*, ed. Farquhar, 306–317.

CHAPMAN, V. J. (1944) "Methods of surveying Laminaria beds" *J. Mar. Biol. Assn. UK* **26**, 37–60.

CHAPMAN, W. M. (1947) "The wealth of the ocean" *Sci. Monthly* NY **44**(3), 192–197.

CHESTERMAN, W., CLYNICK, P. and STRIDE, A. H. (1958) "An acoustic aid to sea bed survey" *Acoustica* **8**, 285–290.

CHESTNUT, A. F. (1950) "The use of a fathometer for surveying shellfish areas" *Science CXI*, **2894**, 677 pp.

CHING, P. A. and WESTON, D. E. (1971) "Wideband studies of shallow water acoustic attenuation due to fish" *J. Sound. Vibr.* **18**(4), 499–810.

CLARKE, G. L. and DENTON, E. J. (1962) "Light and animal life" The Sea Ed. Hill I. 456–468 Interscience, London and New York.

CLARKE, G. L. and WERTHEIM, G. K. (1956) "Measurements of illumination at great depths and at night in the Atlantic Ocean by means of a new bathyphotometer" *Deep Sea Res.* **3**, 189–205.

CLAYDEN, A. D. (1970) Thesis submitted to University of East Anglia.

CRAIG, R. E. (1952) "Echo sounding for herring Part I" *Fishing News*, **2048**, 9–10.

CRAIG, R. E. (1953) "The future of echo detection" *World Fishing* **2.8**, 303–307.

CRAIG, R. E. (1954) "Echo-sounding in marine biology" *Adv. Sci.* **11**, 51–54.

CRAIG, R. E. and FORBES, S. (1969) "Design of a sonar for fish counting" *Fisk. dir. Skrift. Havundersøk.* **15**, 210–219.

CROMWELL, T. (1958) "Thermocline topography, horizontal currents and 'ridging' in the eastern tropical Pacific" *Inter. Amer. Trop. Tuna Comm. Bull.* **3**, 3, 135–162.

CUSHING, D. H. (1951) "The vertical migration of planktonic crustacea" *Biol. Rev.* **26**, 158–192.

CUSHING, D. H. (1952) "Echo surveys for fish" *J. Cons. Perm. Int. Explor. Mer.* **18**, 45–60.

CUSHING, D. H. (1955) "Production and a pelagic fishery" *Fish. Invest. Lond.* **18.7**, 104 pp.

CUSHING, D. H. (1957) "The number of pilchards in the Channel" *Fish. Invest.* 2, **21.5**, 27 pp.

CUSHING, D. H. (1963) "The uses of echo sounding for fishermen" *Fisheries Bull.* HMSO. London, 28 pp.

CUSHING, D. H. (1964) "The counting of fish with an echo sounder" *Rapp, Procès Verb Cons. Explor. Mer.* **155**, 190–194.

CUSHING, D. H. (1966) "The Arctic cod" Pergamon, 93 pp.

CUSHING, D. H. (1968a) "Direct estimation of a fish population acoustically" *J. Fish. Res. Bd. Can.* **25.11**, 2344–2364.

CUSHING, D. H. (1968b) "The abundance of hake off South Africa" *Fish. Invest. Lond.* II, **10**, 20 pp.

CUSHING, D. H. (1969) "The regularity of the spawning season of some fishes" *J. Cons. Perm. Int. Explor. Mer.* **33.1**, 81–92.

CUSHING, D. H. (1969) The use of echo sounders and scanners in the study of fish behaviour" *FAO Fish Rep.* **2**, 115–130.

CUSHING, D. H. (1971) "Upwelling and the production of fish" *Adv. Mar. Biol.* **9**, 255–334.

CUSHING, D. H. (1973) "Computations with the sonar equation" *J. Cons. Int. Explor. Mer.*

CUSHING, D. H. and BRIDGER, J. P. (1966) "The stock of herring in the North Sea and the changes due to fishing" *Fish. Invest. Lond.* 2, **25.1**, 123 pp.

CUSHING, D. H., DEVOLD, F., MARR, J. C. and KRISTJONSSON, H. (1952) "Some modern methods of fish detection" *FAO Fish Bull.* **5**, 3–4, 1–27.

CUSHING, D. H., HARDEN JONES, F. R., MITSON, R. B., PEARCE, G. and ELLIS, G. H. (1963) "Measurements of the target strength of fish" *J. Brit. Instn. Radio Engrs.* **25**, 299–303.

CUSHING, D. H. and HARDEN JONES, F. R. (1967) "Sea trials with modulation sector scanning sonar" *J. Cons. Perm. Int. Explor. Mer.* **30.3**, 324–345.

CUSHING, D. H., LEE, A. J. and RICHARDSON, I. D. (1955) "Echo traces associated with thermoclines" *Sears. Found. J. Mar. Res.* 15, **1**, 1–13.

CUSHING, D. H. and RICHARDSON, I. D. (1955) "A triple frequency echo sounder" *Fish. Invest. Lond.* II, **20.1**, 18 pp.

CUSHING, D. H. and RICHARDSON, I. D. (1955) "Echo sounding experiments on fish" *Fish. Invest. Lond.* II, **18.4**, 34 pp.

CUSHING, D. H. and RICHARDSON, I. D. (1956) "A record of plankton on the echo sounder" *J. Mar. Biol. Ass. U.K.* **35**, 231–240.

DAVIES, D. H. (1957) "The biology of the South African pilchard" Invest. Rep. Div. Fish, S. Africa **32**, 11 pp.

DAVIS, E. E. (1971) "Quasi-synoptic measurements of volume reverberation in the western North Atlantic" *Proc. Int. Symp. Biol. Sound. Scatt. Ocean*, ed. Farquhar, 294–305.

DEVOLD, F. (1951) "Norwegian investigations" *Ann. Biol. Copenh.* **7**, 125–127.

DEVOLD, F. (1952) "A contribution to the study of the migrations of the Atlanto-Scandian herring" *Rapp. Procès. Verb. Cons. Intern. Explor. Mer.* **131**, 103–107.

DEVOLD, F. (1954) "Rapport over tokter for silde under søkelser med 'G.O. Sars' vinteren 1953–1954" *Fiskets Gang* **40**, 261–264 and 275–278.

DEVOLD, F. (1963) "The life history of the Atlanto-Scandian herring" *Rapp. Procès. Verb. Cons. Intern. Explor Mer.* **154**, 98–108.

DEVOLD, F. (1965) "Vintersildinnsigene 1965" *Fiskets Gang* **51**, 378–380.

DEVOLD, F. (1966) "Norwegian adult herring survey 1964" *Ann. Biol. Cons. Intern. Perm. Explor. Mer.* **XXI**, 113–114.

DICKSON, R. R. (1972) "On the relationship between ocean transparency and the depth of some scattering layers in the North Atlantic"

DIETZ, R. S. (1948) "Deep scattering layer in the Pacific and Antarctic Oceans" *J. Mar. Res.* **7.3** 430–442.

DOWD, R. G. (1969) "An echo counting system for demersal fishes" FAO Conf. Fish Behaviour Bergen, *FAO Fish Rep.* **62**, 2, 315–322.

DRAGESUND, O. (1957) "Reactions of fish to artificial light with special reference to large herring and spring herring in Norway" *J. Cons. Perm. Explor. Mer.* **23**, 213–227.

DRAGESUND, O. (1970) "International O group surveys in the Barents Sea" *Co-op. Res. Rep. ICES* A.18, 81 pp.

DRAGESUND, O. and OLSEN, S. (1965) "On the possibility of estimating year class strength by measuring echo abundance of O group fish" *Fisk. Dir. skr. Ser. Havunder sok.* **13(8)**, 47–75.

DREVER, C. and ELLIS, G. H. (1969) "Deep sea trawling and echo sounding techniques" *World Fishing*, 31 pp.

DUVALL, G. E. and CHRISTENSEN, R. J. (1946) "Stratification of sound scatterers in the Ocean" *J. Acoust. Soc. Amer.* **18**, 254.

EDGELL, J. A. (1935) "False echoes in deep water" *Hydrogr. Rev.* **12(1)**, 19–20.

EGGVIN, J. (1934) "Oceanographic conditions at certain Norwegian fishing grounds" *Rapp. Procès. Verb. Cons. Perm. Intern. Mer.* **88**, 4, 11 pp.

ELLIS, G. H. (1956) "Observations on the shoaling behaviour of cod (*Gadus callarias*) in deep water relative to daylight" *J. Mar. Biol. Ass. U.K. N.S.* **35**, 415–417.

ERDMANN, W. (1937) "Die grosse Heringsfischerei in Jahre 1933" *Ber.d. Deutschen Wiss. Komm.* **VIII**, 25–37.

EYRING, C. F., CHRISTENSEN, R. J. and RAITT, R. W. (1948) "Reverberation in the sea" *J. Acoust. Soc. Amer.* **20**, 462.

FEDOROV, S. S. (1960) "The distribution and migrations of immature and young mature Atlantio Scandian herring" *Soviet Invest. North Europ. Seas* 309–329.

FLORES, L. A. and ELIAS, L. A. P. (1967) "Informe preliminar del crucero 6608–9 de invierno 1966 (Mancora-Ilo)" *Infme. Inst. Mar. Peru.* **16**, 24 pp.

FURNESTIN, J. (1953) "Ultrasons et pêche à la sardine au Maroc" *Inst. Pêches. Marit. Maroc Bull.* **1**, 57 pp.

GREER WALKER, M. G., MITSON, R. B. and STORETON WEST, T. (1970) "Trials with a transponding acoustic fish tag tracked with an electronic sector scanning sonar" *Nature* **219.5281**, 196–198.

GULLAND, J. A. (1967) "Analysis of target discrimination with sector scanning equipment" App. to Cushing and Harden Jones (1967).

HANSEN, W. J. and DUNBAR, M. J. (1971) "Biological causes of scattering layers in the Arctic Ocean" *Proc. Int. Symp. Biol. Sound Scatt. Ocean*, ed. Farquhar, 508–526.

HARAM, J. (1965) "Echo sounding in freshwater fishery research" *Proc. Sec. Brit. Coarse Fish Conf. L'Pool* 120–125.

HARDEN JONES, F. R. (1956) "Movements of herring shoals in relation to tidal current" *J. Cons. Perm. Int. Explor. Mer.* 22, **1**, 323–328.

HARDEN JONES, F. R. (1962) "Further observations on the movements of herring (*Clupea harengus*) to tidal current" *J. Cons. Int. Perm. Explor. Mer.* **27.1**, 53–76.

HARDEN JONES, F. R. (1968) "Fish migration" Arnold, London, 325 pp.

HARDEN JONES, F. R. and McCARTNEY, B. S. (1962) "The use of electronic sector scanning sonar for following the movements of fish shoals: sea trials on RRS Discovery II" *J. Cons. Perm. Int. Explor. Mer.* **27**, 141–149.

HARDEN JONES, F. R. and MARSHALL, N. B. (1953) "The structure and functions of the teleostean swim bladder" *Biol. Rev.* **28.1**, 16–23.

HARDEN JONES, F. R. and PEARCE, G. (1958) "Acoustic reflexion experiments with perch (*Perca fluviatilis* Linn.) to determine the proportion of the echo returned by the swim bladder" *J. exp. Biol.* **35**, 437–450.

HARDY, A. C. (1924) "Report on the possibilities of aerial spotting of fish" *Fish. Invest. Lond.* II, **7.5**, 8 pp.

HARDY, A. C. and GUNTHER, E. R. (1935) "The plankton of the South Georgia whaling grounds" *Discovery Rep.* **XI**, 1–456.

HARDY, A. C., HENDERSON, G. T. D., LUCAS, C. E. and FRASER, J. H. (1936) "The ecological relation between the herring and the plankton investigated with the plankton indicator" *J. Mar. Biol. Assn. U.K. NS* **21**, 147–291.

HART, T. J. and CURRIE, R. I. (1960) "The Benguela current" *Disc. Rep.* **31**, 1–297.

HASHIMOTO, T. (1953) "Characteristics of ultrasonic waves transmitted vertically in the water" *Sci. Rep. Fish. Boat Lab.* **2**, 71–167.

HASHIMOTO, T. and MANIWA, Y. (1956) "Study of DSL by ultrasonic wave (1)" *J. Tokyo Univ. Fish.* **42**, 2, 113–123.

HASHIMOTO, T. and MANIWA, Y. (1956) "Study on reflection loss of ultrasonic wave on fish body by millimeter wave" *Tech. Rep. Fish. Boat Lab.* **8**, 113–118, Tokyo.

HASHIMOTO, T. and MANIWA, Y. (1958) "Study on ultrasonic reflection loss on fish body (on the influence of the airbladder)" *Tech. Rep. Fish. Boat Lab.* **11**, 143–155.

HASLER, A. D. and VILLEMONTE, J. (1953) "Observations on the daily movements of fishes" *Science* **118**, 321–322.

HASLETT, R. W. G. (1962a) Measurement of the dimensions of fish to facilitate calculations of echo-strength in acoustic fish detection" *J. Cons. Int. Explor. Mer.* **27**, 261–269.

HASLETT, R. W. G. (1962b) "The back-scattering of acoustic waves in water by an obstacle II: Determination of the reflectivities of solids using small specimens" *Proc. Phys. Soc.* **79**, 559–571.

HASLETT, R. W. G. (1962c) "Determination of the acoustic scatter patterns and cross-sections of fish models and ellipsoids" *Br. J. Appl. Phys.* **13**, 611–620.

HASLETT, R. W. G. (1962d) "Determination of the acoustic back-scattering patterns and cross sections of fish" *Br. J. Appl. Phys.* **13**, 249–357.

HEARN, P. T. (1970) "A fish counter for use in commercial fishing" FAO Tech. Conf. Fish Finding, Purse Seining and Aimed Trawling, Reykjavik 1970, Paper 54, 5 p.

HENTSCHEL, E. and WATTENBERG, H. (1930) "Plankton und Phosphat in der Oberflächenschicht des Südatlantischen Ozeans" Ann. Hydrogr. Berl. **58**, 273–277.

HERMANN, F. (1952) "Hydrographic conditions in the south western part of the Norwegian sea" Ann. Biol. Cons. Perm. Intern. Explor. Mer. **VII**, (1951), 23–26.

HERSEY, J. B. and BACKUS, R. H. (1954) "New evidence that migrating gas bubbles, probably the swim bladders of fish, are largely responsible for scattering layers in the continental rise south of New England" Deep Sea Res. **1**, 190–191.

HERSEY, J. B. and BACKUS, R. H. (1962) "Sound scattering by marine organisms" The Sea Ed. Hill Vol. 1, 498–539, Interscience.

HERSEY, J. B., BACKUS, R. H. and HELLWIG, J. (1962) "Sound scattering spectra of deep scattering layers in the western north Atlantic ocean" Deep Sea Res. **8**, 314, 196–210.

HERSEY, J. B., EDGERTON, H. E., RAYMOND, S. O. and HAYWARD, G. (1961) "Pingers and thumpers advance deep sea exploration" J. Instrument. Soc. Amer. **87**, 277 pp.

HERSEY, J. B., JOHNSON, H. R. and DAVIS, L. C. (1952) "Recent findings about the deep scattering layer" J. Mar. Res. **11**, 1–9.

HERSEY, J. B. and MOORE, H. B. (1948) "Progress report on scattering layer observations in the Atlantic ocean" Trans. Amer. Geophys. Un. **29**, 341–354.

HESTER, F. J. (1967) "Identification of biological sonar targets from body motion Doppler shifts" Marine Bioacoustics **2**, 59–74, Pergamon.

HICKLING, C. F. (1927) "The natural history of the hake I and II" Fish. Invest. Lond. 2, **10**, 100 pp.

HJORT, J. and RUUD, J. (1929) "Whaling and fishing in the North Atlantic" Rapp. Procès. Verb. Cons. Int. Perm. Explor. Mer. 56, **1**, 123 pp.

HODGSON, W. C. (1950) "Echo sounding and the pelagic fisheries" Fish. Invest. Lond. II **17.4**, 25 pp.

HODGSON, W. C. (1957) "The herring and its fishery" Routledge and Kegan Paul, London, 197 pp.

HODGSON, W. C. and FRIDRIKSSON, A. (1955) "Report on echo sounding and Asdic for fishing purposes" Rapp. Procès-Vert. Cons. Int. Explor. Mer. **139**, 49 pp.

HODGSON, W. C. and RICHARDSON, I. D. (1949) "The experiments on the Cornish pilchard fishery, 1947–48" Fish. Invest. Lond. II 17, **2**, 21 pp.

HOFFMAN, J. (1957) "Hyperbolic curves applied to echo sounding" Intern. Hydrogr. Rev. **34.2**, 45–56.

HØGLUND, H. (1955) "Swedish herring tagging experiments 1949–1953" Rapp. Procès-Verb. Cons. Intern. Explor. Mer. **140.2**, 19–29.

HØGLUND, H. (1960) "On the herring in Bohuslän during the Great Fishing Period of the Eighteenth Century" Int. Council Explor. Sea 1960 paper 1925.

HOLMES, R. W., SCHAEFER, M. and SHIMADA, B. M. (1957) "Primary production, chlorophyll and zooplankton volumes in the tropical eastern Pacific ocean" Bull. Inter-Amer. Trop. Tuna Comm. II. **4**, 129–169.

HOURSTON, A. L. (1953) "The spawning population of herring in northern British Columbia in 1952" Progr. Rep. Pac. Coast Stas. **97**, 21–27.

HUNKING, K. (1965) "The seasonal variation in the sound scattering layer observed at Fletcher's Ice Island (T.3) with a 12 khz echo sounder" Deep Sea Res. **12**, 879–881.

HYLEN, A., MIDTTUN, L. and SAETERSDAL, G. (1961) "Torskeundersøkelsene i Lofoten og i Barents Havet 1960" Fisken og Havet. **2**, 1–14.

JAKOBSSON, J. (1963) "Some remarks on the distribution and availability of the Iceland north coast herring" Rapp. Procès. Verb. Cons. Perm. Intern. Explor. Mer. **154**, 73–83.

JOHNSON, H. R., BACKUS, R. H., HERSEY, J. B. and OWEN, D. M. (1956) "Suspended echo-sounder and camera studies of midwater sound scatterers" Deep Sea Res. **3**, 266–272.

JOHNSON, M. W. (1948) "Sound as a tool in marine ecology from data on biological noises and the deep scattering layer" J. Mar. Res. **7**, 443–458.

JOHNSON, W. E. (1961) "Aspects of the ecology of a pelagic zooplankton eating fish" Verh. int. Verein. Theor. angew. Limnol. **14**, 727–731.

KAMPA, E. M. and BODEN, B. P. (1954) "Submarine illuminations and the twilight movements of a sonic scattering layer" Nature **174**, 4436, 869–871.

KAMPA, E. M. and BODEN, B. P. (1957) "Light generation in a sonic scattering layer" Deep Sea Res. **74**, 73–92.

KANWISHER, J. and VOLKMANN, G. (1955) "A scattering layer observation" Science **121**, 108–109.

KING, J. E. and HIDA, T. S. (1954) "Zooplankton abundance in the central Pacific" U.S. Fish and Wildlife Service. Fish. Bull. **57.118**, 365–395.

KINZER, J. (1969) "On the quantitative distribution of zooplankton in deep scattering layers" Deep Sea Res. **16.2**, 117–126.

KOMAKI, Y. and MATSUYE, Y. (1956) "Observations on the so-called Deep Scattering Layers (DSL) with special reference to the vertical distribution of plankton" J. Tokyo Univ. Fish. **42.2**, 139–150.

KONSTANTINOV, K. G. (1964a) "Water temperature as a factor guiding fishes during their migrations" ICNAF Spec. Publ. 6, ICNAF Environmental Symp. 221–225.

KONSTANTINOV, K. G. (1964b) "Diurnal, vertical migrations of demersal fish and their possible influence and the estimation of fish stocks" *Intern. Rapp. Procès-Verb. Cons. Explor. Mer.* **155**, 23–26.

KOCSY, F. (1954) "A survey of the deep sea features taken during the Swedish deep sea expedition" *Deep Sea Res.* **1.3**, 176–184.

KREFFT, G. and SCHUBERT, K. (1950) "Beobachtungen über Einfluss künstlicher Beleuchtung der Meeresoberfläche auf Fische" Fischereiwelt 1950, **6**, 86–88.

KUROKI, T. (1969) "Problems of designing fishing gear as related to the behaviour of fish" FAO Conf. Fish Behaviour, *FAO Fish Rep.* **62.2**, 523–542.

LEBEDEVA, L. P. (1965) "Measurement of the dynamic complex shear modulus"

LEE, A. J. (1952) "The influence of hydrography on the Bear Island cod fishery" *Rapp. Procès. Verb. Cons. Intern. Explor. Mer.* **131**, 74–102.

LEGALL, J. (1952) "La détection des bancs de poissons" *Rapp. Procès. Verb. Cons. Int. Explor. Mer.* **132**, 65–71.

LENZ, J. (1965) "Zur Ursache der an die Sprungschicht gebundenen Echostreuschichten in der westlichen Ostsee" *Ber. d. Wiss. Komm. Meeres.* **18.2**, 111–161.

LOVE, R. H. (1971a) "Measurements of fish target strength: a review" *Fish. Bull. US.* NOAA **69.4**, 703–715.

LOVE, R. H. (1971b) "Dorsal aspect target strength of an individual fish" *J. Acoust. Soc. Amer.* **49.3**, 816–823.

LUCAS, C. E. (1935) "On the diurnal variation of size groups of trawl-caught herring" *J. Cons. Int. Explor. Mer.* **XI.1**, 53–59.

LUNDBECK, J. (1953) "Biologisch—statistischer Bericht über die deutsche Hochseefischerei in Jahre 1952" Jahresbericht uber Deutsche Fischerei 1952, 146–182, Berlin.

LUNDBECK, J. (1954) "Biologisch—statistischer Bericht über die deutsche Hochseefischerei in Jahre 1953" Jahresbericht uber Deutsche Fischerei 1953, 135–171, Berlin.

LUNDBECK, J. (1955) "Biologisch—statistischer Bericht über die deutsche Hochsee fischerei in Jahre 1954" Jahresbericht uber Deutsche Fischerei 1954, 143–172, Berlin.

McCARTNEY, B. S. and STUBBS, A. R. (1971) "Measurements of the acoustic target strengths of fish in dorsal aspect, including swim bladder resonance" *J. Sound. Vibr.* **15.3**, 397–420.

McCARTNEY, B., STUBBS, A. R. and TUCKER, M. J. (1965) "Low frequency target strength of pilchard shoals and the hypothesis of swimbladder resonance" *Nature* **207**, 4992, 39–40.

MACHLUP, S. and HERSEY, J. B. (1955) "Analysis of sound scattering observations from non-uniform distributions of scatterers in the Ocean" *Deep Sea Res.* **3**, 1–22.

MAHROUS, M. A. and CUSHING, D. H. (1956) "Acoustic measurements on fish" Third Arab Science Congress, 750–756.

MARGETTS, A. R. (1967) "Catching clupeoid fish by pelagic trawl" FAO Conf. Fish Behaviour, *FAO Fish Rep.* **62.2**, 499–508.

MARSHALL, N. B. (1951) "Bathypelagic fishes as sound scatterers in the ocean" *J. Mar. Res.* **10**, 1–17.

MARSHALL, N. B. (1961) "Swim bladder structure in relation to the deep sea fish systematics" *Discovery Res.* **31**, 1–222.

MARSHALL, N. B. (1971) "Swim bladder development and the life of deep sea fishes" *Proc. Int. Symp. Biol. Sound. Scatt. Ocean,* ed. Farquhar, 69–73.

MARSHALL, P. T. (1958) "Primary production in the Arctic" *J. Cons. Int. Explor. Mer.* **23.2**, 173–177.

MARTY, J. J. and WILSON, A. P. (1960) "Migrations of the Atlanto-Scandian herring" Soviet Invest. North Europ. Seas 329–341.

MAUCORPS, A. (1966) "La campagne harenguière 1964–1965 dans la région du Pas-de-Calais" *Ann. Biol. Cons. Perm. Intern. Explor. Mer.* **XX1**, 160–161.

MIDTTUN, L. (1964) "The relation between temperature conditions and fish distribution in the south east Barents Sea" ICNAF Spec. Publ. 6. ICNAF Environmental Symp. 213–221.

MIDTTUN, L. (1966) "Note on the measurement of target strength of fish at sea" ICES 1966 mimeo doc F **9**, 2 p.

MIDTTUN, L. and HOFF, L. (1962) "Determination of the reflection of sound by fish" *Fiskeridir. Skr. Havundersok.* **13(3)**, 18 pp.

MIDTTUN, L. and NAKKEN, O. (1971) "On acoustic identification, sizing and abundance of fish" *Fisk. Dir. Skrift. Havundersøk.* **16.1**, 36–48.

MIDTTUN, L. and SAETERSDAL, G. (1957) "On the use of echo sounder observations for estimating fish abundance" ICNAF Spec. Publ. 2 (Lisbon meeting) Paper 29, 2 p.

MITCHELL, J. (1864) "The herring, its natural history and national importance" Edinburgh and London.

MITSON, R. B. (1969) "The structure and functioning of sonar equipment" *FAO Fish. Tech. Paper* **83**, 17–29.

MITSON, R. B. and COOK, J. C. (1970) "Shipboard installation and trials of an electronic sector scanning sonar" *Radio and Electr. Engin.* **41.8**, 339–350.

Mitson, R. B. and Wood, R. J. (1962) "An automatic method of counting fish echoes I. The equipment, II. Trials with the fish counter" *J. Cons. Int. Explor. Mer.* **26(3)**, 281–291.

Mohr, H. (1963) "Zum Verhalten von Fischschwarmen beim Fang mit pelagischen Netzen" Protokolle Fischereitechnik VIII 149–159.

Monstad, T., Nakken, O. and Naevdal, G. (1969) "Skreiinnsiget 1969" *Fiskets Gang* 55 **34**, 571–573.

Moore, H. B. (1950) "The relation between the scattering layer and the Euphausiacea" *Biol. Bull* **99**, 181–212.

Motoda, S. and Hirano, Y. (1963) "Review of Japenese herring investigations" *Rapp. Proc. Verb. Cons. Int. Perm. Mer.* **154**, 249–262.

Murphy, G. I. and Shomura, R. (1955) "Long line fishing for deep swimming tunas in the Central Pacific, August–November 1952" *U.S. Fish and Wildlife Service Spec. Sci. Rep.* **137**, 42 pp.

Nakamura, H. (1959) "Average year's fishing condition of tuna long line fisheries" Fed. Jap. Tuna Fishermen's Co-op Assoc. Transl. van Campen (1962) BCF Honolulu.

Nakamura, H. (1969) "Tuna distribution and migration" Fishing News (Books), London, 76 pp.

Nikoronov, I. V. (1959) "The basic principles of fishing for the Caspian kilka by underwater light" Modern Fishing Gear of the World, ed. Kristjonsson, Fishing News (Books) Ltd. 559–566.

Nishimura, M. (1961) "Study on echo sounder for tuna fishing in the north east New Zealand sea" *Tech. Rep. Fish. Boat* **15**, 91–109.

Nishimura, M. (1963) "Investigation of tuna behaviour by fish finder" *FAO Fish Rep.* **6**, 1113–1123.

Northcote, T. G. (1964) "Use of a high frequency echo sounder to record distribution and migration of *Chaoborus* larvae" *Limmol. Oceanogr.* **9.1**, 87–89.

Ogura, M. (1968) "Studies on mackerel angling fisheries III. The effect of fish finders on the catch of the mackerel vertical long line fishery" *Bull. Jap. Soc. Sci. Fish.* **34.7**, 576–580.

Okonski, S. (1969) "Echo sounding observations of fish behaviour in the proximity of the trawl" FAO Co-op Fish Behaviour, *FAO Fish. Rep.* **62.2**, 377–388.

Olsen, S. (1960) "Observations on sound scatterers in Newfoundland waters" *J. Fish. Res. Bd. Can.* **17(2)**, 211–219.

Olsen, S. (1969) "A note on estimating school size from echo traces" *FAO Fish Rep.* **78**, 37–40.

Olson, F. C. W. (1964) "The survival value of fish schooling" *J. Cons. Int. Perm. Explor. Mer.* **29**, 1915.

Otsu, T. (1960) "Albacore migration and growth in the North Pacific ocean as estimated from a tag recoveries" *Pacific Sci.* **14(3)**, 257–266.

Parrish, B. B. (1952) "Echo sounding for herring—Part 2" *Fishing News Lond.* **2055**, 7–10.

Pavshtiks, E. A. (1960) "Main regularities in the plankton development in the Norwegian and Greenland Seas" Soviet Fish. Invest. in North European Seas, 155–172.

Pearce, G. and Philpott, S. (1970) "A fish detection echo sounding system with an electronically stabilized narrow beam and bottom locked display" FAO Tech. Conf. Fish Finding Purse Seining and Aimed Trawling. Paper 45.

Percier, R. (1953) "Etudes des bancs de poissons et surveillance des lieux de pêche au moyens des sondeurs ultrasons" *Pêche marit.* **905**, 344–347.

Pickwell, G. V., Barham, E. G. and Wilton, J. W. (1964) "Carbon monoxide production by a bathypelagic siphonophore" *Science* **144**, 3620, 860–862.

Posner, G. S. (1957) "The Peru current" *Bull. Bingh. Oceanogr. Colln.* **16(2)**, 106–155.

Postuma, K. H. (1957) "The vertical migration of feeding herring in relation to light and the vertical temperature gradient" ICES Mimeo 1957.

Raitt, R. W. (1948) "Sound scatterers in the sea" *J. Mar. Res.* **7.3**, 393–409.

Reid, J. L. (1962) "On circulation, phosphate-phosphorus and zooplankton volumes in the upper part of the Pacific Ocean" *Limnol. Oceanogr.* **7.3**, 287–306.

Renou, J. and Tchernia, P. (1947) "Détection des bancs de poissons par ultrasons" Min. Marine des Côtes Comité d'Océanogr et d'Études 21–29.

Richardson, I. D. (1950) "The use of an echo sounder to chart sprat concentrations" *Ann. Biol. Cons. Int. Explor. Mer.* **6**, 136–139.

Richardson, I. D. (1951) "Echo sounder surveys for sprats in the 1950–51 seasons" *Ann. Biol. Cons. Int. Explor. Mer.* **7**, 97–98.

Richardson, I. D. (1952) "Some reactions of pelagic fish to light as recorded by echo sounding" *Fish. Invest. Lond.* II **18.1**, 20 pp.

Richardson, I. D., Cushing, D. H., Harden Jones, F. R., Beverton, R. J. and Blacker, R. W. (1959) "Echo sounding experiments in the Barents Sea" *Fish. Invest. Lond.* **22.9**, 55 pp.

Riley, G. A., Stommel, H. and Bumpus, D. F. (1949) "Quantitative ecology of the plankton of the western North Atlantic" *Bull. Bingh. Oceanogr. Coll.* **12.3**, 1–169.

Rollefsen, G. (1955) "Observations on the cod and cod fisheries of Lofoten" *Rapp. Procès. Verb. Cons. Intern. Explor. Mer.* **136**, 40–47.

Runnstrøm, S. (1936) "A study on the life history and migrations of the Norwegian spring herring based on the analysis of the winter rings and summer zones of the scale" *Fisk. Dir. Skr. Ser. Havunders* **5.2**, 103 pp.

RUNNSTRØM, S. (1941) "Racial analysis of the herring in Norwegian waters" *Fiskeridir. Skr. Havundersok* **6(7)**, 1–110.

RUSBY, J. S. M., DOBSON, R., EDGE, R. H., PIERCE, F. E. and SOMERS, M. L. (1969) "Records obtained from the trials of a Long Range Side Scan Sonar (GLORIA project)" *Nature* **223**, 1225–1227.

RUSSELL, F. S. (1928) "The vertical distribution of marine macroplankton. VII. Observations on the behaviour of *Calanus finmarchicus*" *J. Mar. Biol. Assn.* U.K IV5. **XV.2**, 429–459.

SAETERSDAL, G. (1960) "Ekkolodd fiskesoking og untervannsfotografiering" *Konkylien* **5.1**, 28–31.

SAETERSDAL, G. and HYLEN, A. (1959) "Skreiundersøkelsene og skreifisket i 1959" *Fisken og Havet* **1**, (Dec) 1–18.

SAILA, S. and FLOWERS, J. M. (1967) "Elementary applications of search theory to fishing tactics as related to some aspects of fish behaviour" FAO Conf. Fish Behaviour, *FAO Fish Rep.* **62.2**, 343–356.

SANO, N. (1968) "On some techniques of detecting salmon by the echo sounding method I Estimation of the fishing grounds of salmon according to swimming speed and swimming depth calculated from echo traces on recording paper" *Bull. Jap. Soc. Sci. Fish.* **34.8**, 660–669.

SARS, G. O. (1864–9) "Reports on the Lofoten cod fisheries" Christiania.

SAVAGE, R. E. (1930) "The influence of *Phaeocystis* on the migration of the herring" *Fish. Invest. Lond.* II. **XII.2**, 14 pp.

SAVAGE, R. E. (1931) "The relation between the feeding of the herring off the east coast of England and the plankton of the surrounding waters" *Fish. Invest. Lond.* II. **XII.2**, 88 pp.

SAVAGE, R. E. (1937) "The food of the North Sea herring" *Fish. Invest. Lond.* II. **15.5**, 60 pp.

SCHAEFER, M. B., BISHOP, Y. M. M. and HOWARD, G. (1958) "Some aspects of upwelling in the Gulf of Panama" *Bull. Inter. Am. Trop. Tuna Comm* **3(2)**, 77–131.

SCHÄRFE, J. (1960) "Fortsetzung der Einschiff-Swimmschleppnetz-Versuche mit dem FD 'Rendsburg' in der Zeit von 12 bis 27.9.1959 in der Nordsee" *Protokolle Fischereitechnik* VI 158–189.

SCHÄRFE, J. (1968) "Das deutsche Einschiff-Schwimmschleppnetz" *Inform. Fischwirts.* **15**, 314, 104–167.

SCHÄRFE, J. and STEINBERG, R. (1963) "Neue Erfahrungen mit Schwimn schleppnretzen (Juli bis Dez 1962)" *Protokolle Fischereitechnik*, VIII 160–230.

SCHMIDT, U. (1957) "Beitrage zur Biologie des Kohlers (*Gadus virens* L.) in den Island fischen Gewässern" *Ber. d. Komm. Wiss. Meeres.* **14**, 46–82.

SCHRÖDER, H. and SCHRÖDER, R. (1964) "On the use of the echo sounder in lake investigations" *Mem. 1st Ital. Idrobiol. Dot. Marco de Marchi* **17**, 167–188.

SCHUBERT, K. (1958) "Survey of the German Commercial fisheries and the biological conditions of the stock" *Ann. Biol. Cons. Perm. Intern. Explor. Mer.* **XXIII**, 192–197.

SCHÜLER, F. (1951) "Die abbildungstreue von Meeresboden-Profilen in Echogrammen" *Dtsch. Hydrogr. Z.* **4**, 97–105.

SCHÜLER, F. and KREFFT, G. (1951) "Zur Frage der Verwendung des Echographen in der Loggerfischerei" *Fischeriwelt*, **3.4**, 63 pp.

SCOFIELD, W. L. (1929) "Sardine fishing methods at Monterey, California" *Fish. Bull. Sacramento* **19**, 61 pp.

SHIBATA, K. (1962) "Analysis offish finding records I" *Bull. Fac. Fish. Nagasaki Univ.* **13**, 9–18.

SHIBATA, K. (1963) "Analysis of fish finding records IV. Report on the Deep Scattering Layer and tuna food" *Bull. Fac. Fish. Nagasaki Univ.* **15**, 59–84.

SHIBATA, K. (1968) "Analysis of fish finding records VII. Swimming speeds of fish" *Bull. Fac.Fish. Nagasaki Univ.* **25**, 59–66.

SHIBATA, K. (1970) "Study on details of ultrasonic reflection from individual fish" *Bull. Fac. Fish. Nagasaki Univ.* **29**, 1–82.

SHIBATA, K., NISHIMURA, K., AOYAMA, T. and YAMANAKA, I. (1970) "Development of echo counting system for estimating fish stocks in Japan" FAO Tech. Conf. Fish Finding, Purse Seining and Aimed Trawling. Paper 38.

SIMPSON, A. C. (1959) "The spawning of plaice in the North Sea" *Fish. Invest. Lond.* II **22(7)**, 100 pp.

SMITH, O. R. (1947) "The location of sardine schools by supersonic echo ranging" *Comm. Fish. Rev.* **9**, 1.

SMITH, O. R. and AHLSTROM, E. H. (1948) "Echo ranging for fish schools and observations on temperature and plankton off Central California in the spring of 1941" *Spec. Sci. U.S. Fish and Wildlife Serv.* **44**, 1–29.

SMITH, P. E. (1970) "The horizontal dimensions and abundance of fish schools in the upper mixed layer as measured by sonar" *Proc. Intnl. Symp. Biol. Scatt. Ocean*, ed. Farquhar, 563–591.

SMITH, P. F. (1954) "Further measurements of the sound properties of several marine organisms" *Deep Sea Res.* **2**, 71–79.

SPRAGUE, L. M. and VROOMAN, A. M. (1963) "A racial analysis of the Pacific sardine (*Sardinops caerulea*) based on the study of erythrocyte antigens" *Ann. N.Y. Acad. Sci.* **97**, 1, 131–138.

STEELE, J. H. (1956) "Plant production in the northern North Sea" *Mar. Res. Scot.* **7**, 366 pp.

STEELE, J. H. (1961) "The environment of a herring fishery" *Marine Res. Scot.* **6**, 1–19.

STEINBERG, R. (1960) "Einschiff-Schwimmschleppnetz-Versuche mit einem Logger vom 17.11 bis 3.12.59 in der Irischden See und in Kanal" *Protokolle Fischereitechnik* VI, 235–253.

STEINBERG, R. (1967) "Entwicklung und gegenwartiger Stand der pelagischen Schleppnetz Fischerei der deutschen Kombilogger" *Protokolle Fischereitechnik* X, **47**, 213–318.

SUDA, A. (1963) "Structure of the albacore stock and fluctuation in the catch in the North Pacific area" Proc. World Sci. Meeting Biology Tunas 1962, 1237–1277.

SUDA, A., KOTO, R. and KUMES, S. (1963) "An outline of the tuna longline grounds in the Indo Pacific" *FAO Fish. Rep.* 6, **6**, 1163–1176.

SUND, O. (1932) "On the German and Norwegian observations on the cod in 1931" *Rapp. Procès. Verb. Cons. Intern. Explor. Mer.* **81**, 151–156.

SUND, O. (1935) "Echo sounding in fishery research" *Nature* **135**, 3423, 953.

SUND, O. (1935) "Lofotfisket 1935" *Arsberetning vedk. Norges Fisk.* 1935 **2**, 89–91.

SUND, O. (1943) "The fat and small herring on the coast of Norway in 1940" *Ann. Biol. Cons. Int. Explor. Mer.* **1**, 58–79.

SUNDNES, G. (1964) "Om skreiens at ferd på gytefeltet" *Fisken og Havet*, **4**, 1964, 1–5.

SUOMALA, J. B., JR. (1970) "The application of a digital computer simulation to aid in the evaluation of echo sounder design and performance" FAO Tech. Conf. Fish Finding, Purse Seining and Aimed Trawling, Paper 29.

TÅNING, A. VEDEL, EINARSSON, H. and EGGVIN, H. (1957) "Records from the month of June of the Norwegian-Icelandic herring stock in the open ocean" *Ann. Biol. Cons. Perm. Intern. Explor. Mer.* XII, 165–166.

TCHERNIA, P. (1949–50) "Observations d'océanographie biologique faites par l'aviso polaire 'Commandant Charcot' pendant la campagne 1948–9" *Bull. Inf. Com. Centr. Océanographie et d'Etudes des Côtes* I.

TCHERNIA, P. (1951) "Compte-rendu préliminaire des observations océanographiques faites par le batiment polaire 'Commandent Charcot' pendant la campagne 1949–50" *Bull. Inf. Com. Centr. Océanographique et d'Etudes des Côtes* III, **2**, 40–57.

TESTER, A. L. (1943a) "Locating herring with an echo sounder" *Progr. Rep. Pac. Coast Stas. Fish. Res. Bd. Can.* **55**, 4–6.

TESTER, A. L. (1944) "Echosounding for summer herring" *Progr. Rep. Pac. Coast Sta. Fish. Res. Bd. Can.* **61**, 17–20.

THOMPSON, D'ARCY (1947) "A glossary of greek fishes" Oxford Univ. Press.

THOMPSON, H. (1929) "Haddock Biology" *Rapp. Procès. Verb. Cons. Intern. Explor. Mer.* **54**, 135–163.

THORNE, R. E., REEVES, J. E. and MILLIKAN, A. E. (1971) "Estimation of the Pacific hake (*Merluccius productus*) population in Port Susan, Washington, using an echo integrator" *J. Fish. Res. Bd. Can.* **28**, 9, 1275–1284.

THORNE, R. E. and WOODEY, J. C. (1970) "Stock assessment by echo integration and its application to juvenile sockeye salmon in Lake Washington" *Washington Sea Grant Publ.* **70–2**, 31 pp.

TOWNSEND, C. H. (1935) "The distribution of certain whales as shown by log book records of American whale ships" *Zoologica* **19**, 1–50.

TROUT, G. C. (1957) "The Bear Island cod; migration and movements" *Fish. Invest. Lond.* II. **21**, 6 pp.

TROUT, G. C., LEE, A. J., RICHARDSON, I. D. and HARDEN JONES, F. R. (1952) "Recent echo sounder studies" *Nature* **170**, 4315, 71–72.

TRUSKANOV, M. D. and SCHERBINO, M. (1963) "The determination of the size of fish shoals using hydroacoustic instruments" *Rybnoe Khoz.* **6**, 52–58.

TRUSKANOV, M. and SCHERBINO, M. (1964) "On determination of volumes of dense concentrations of Atlantic and Scandinavian herring" *Trudy PINRO* **14**, 183–202.

TRUSKANOV, M. and SCHERBINO, M. (1966) "Methods of direct calculation of fish concentrations by means of hydroacoustic apparatus" *ICNAF Res. Bull.* **3**, 70–89.

TRUSKANOV, M. and SCHERBINO, M. (1966) "Methods of direct calculation of fish concentrations by means of hydroacoustic apparatus" *ICNAF Res. Bull.* **3**, 70–80.

TUCKER, G. H. (1951) "Relation of fishes and other organisms to the scattering of underwater sound" *J. Mar. Res.* **10**, 215–238.

TUCKER, D. G. and WELSBY, V. G. (1960) "Electronic sector-scanning asdic; an improved fish locator and navigational instrument" *Nature* **185**, (4709) 277–279.

TUCKER, D. G. and WELSBY, V. G. (1964) "Sector scanning sonar for fisheries purposes" *Mod. Fish. Gear World* **2**, 367–371.

TUNGATE, D. S. (1958) "Echo sounder surveys in the autumn of 1956" *Fish. Invest. Lond.* II **22.2**, 17 pp.

TVEITE, S. (1969) "Zooplankton and the discontinuity layer in relation to echo traces in the Oslo fjord" *Fisk. Dir. Hav.* **15.2**, 25–35.

UDA, M. (1937) "Researches on 'Siome' or current rip in the seas and oceans" *Geophys. Mag.* XI. **4**, 307–372.

UDA, M. (1952) "On the relation between the variation of the important fisheries conditions and the oceanographical conditions in the adjacent waters of Japan" *J. Tokyo Univ. Fish.* **38.3**, 363–389.

UDA, M. (1956) "Researches on the fishing grounds in relation to the scattering layer of supersonic waves" *J. Tokyo Univ. Fish.* **42.2**, 103–111.

URICK, R. J. (1967) "Principles of underwater sound for engineers" McGraw Hill, 342 pp.

VALDEZ, V. (1961) "Echo sounder records of ultrasonic sounds made by killer whales and dolphins" *Deep Sea Res.* **7**, 289–290.

VALDEZ, V. and CUSHING, D. H. (1962) "The effect of tidal streams on the presence of an extensive layer of midwater echo traces" *J. Cons. Int. Perm. Explor. Mer.* **27.3**, 236–247.

VALDEZ, V. and CUSHING, D. H. (1966) "The diurnal variation in depth and quantity of echo traces and their distribution in area in the Southern Bight of the North sea" *J. Cons. Intern. Perm. Explor. Mer.* **30.2**, 237–255.

VAN CAMPEN, W. G. (1960) "Japanese summer fishery for albacore (*Germo alalumga*)" *U.S. Fish and Wildlife Service Res. Rep.* **52**, 1–29.

VAN SCHUYLER, P. (1971) "An acoustically determined distribution of resonant scattering north of Oahu" *Proc. Int. Symp. Biol. Sound. Scatt. Ocean*, ed. Farquhar, 328–340.

DE VEEN, J. F. (1964) "On the phenomenon of soles swimming near the surface of the sea" *Rapp. Procès. Verb. Cons. Int. Explor. Mer.* **155**, 51.

VERHEIJEN, F. J. (1958) "The mechanisms of the trapping effect of artificial light sources upon animals" *Arch. neèrl Zool.* **13**, 1–107.

VERHEIJEN, F. J. (1959) "Attraction of fish by the use of light" *Modern Fish. Gear*, World I, 548–549.

VESTNES, G., STROM, A., SAETERSDAL, G. and VILLEGAS, L. (1965) "Informs sobre investigationes exploratiores en la zona de Talcahuano, Valdivia y Puerto Mont, Junio-Julio 1965, realizada con el B/1" Carlos Darwin Inst. Fomento Pesquero 10.

VOGLIS, G. M. and COOK, J. C. (1966) "Underwater applications of an advanced acoustic scanning equipment" *Ultrasonics* **4**, 1–9.

V. BRANDT, A. (1960) "Einschiff Schwimmschleppnetz—Versuche mit den F. D. RENDSBURG in der Zeit vom 28.10 bis 12.11.59 in der Irischen See und in Kanal" *Protokolle Fischereitechnick* VI, 190–213.

V. BRANDT, A. and SCHARFE, J. (1950) "Zur quantitativen Auswertung der Echolotbeobachtungen" *Protokolle Fischereitechik* **2**, 4 p.

WELSBY, V. G., BLAXTER, J. H. S. and CHAPMAN, C. J. (1963) "Electronically scanned sonar in the investigation of fish behaviour" *Nature* **199** (4897), 980–981.

WELSBY, V. G., DUNN, J. R., CHAPMAN, C. J., SHARMAN, D. P. and PRIESTLEY, R. (1964) "Further uses of electronically scanned sonar in the investigation of behaviour of fish" *Nature* **203**, 4945, 588–589.

WESTENBERG, J. (1963) "Acoustical aspects of some Indonesian fisheries" *J. Cons. Int. Explor. Mer.* **18.3**, 311–325.

WESTON, D. E. (1958) "Observations on a scattering layer at the thermocline "*Deep Sea Res.* **5**, 44–50.

WESTON, D. E. (1967) "Sound propagation in the presence of bladder fish" in VM Albers (ed) *Underwater Acoustics*, **2**, Ch. 5, 55–88.

WESTON, D. E., HORRIGAN, A. A., THOMAS, S. J. L. and REVIE, J. (1969) "Studies of sound transmission fluctuations in shallow coastal waters" *Phil. Trans. Roy. Soc.* **265**, 567–608.

WESTON, D. E. and REVIE, J. (1971) "Fish echoes on a long range sonar display" *J. Sound Vibr.* **17.0**, 105–112.

WIBORG, K. V. (1960) "Investigations on pelagic fry of cod and haddock in coastal and offshore areas of northern Norway in July-August 1957" *Fiskeridir. Skr. Ser. Havundersog.* **12**, 8.

WILCOCKS, J. (1882) "The history and statistics of the pilchard fishery in England" Fish and Fisheries, 290–303, Blackwood.

WILLIAMS, F. (1968) "Report on the Guinean trawling survey" *Org. Afr. Unity Sci. Tech.* I 828 pp. II 529 pp.

WOOD, A. B., SMITH, F. D. and McGEACHY, J. A., British Patent 375375 and (1935) J.I.E.E. 76, 550.

WOODHEAD, A. D. and WOODHEAD, P. M. J. (1959) "The effects of low temperature on the physiology and distribution of the cod, *Gadus morhua* L. in the Barents Sea" *Proc. Zool. Soc. Lond.* **133(2)**, 181–199.

WOOSTER, W. S. and CROMWELL, T. (1958) "An oceanographic description of the eastern tropical Pacific" *Bull. Scripps. Inst. Oceanogr.* **7.3**, 169–282.

YABE, H., YABUTA, Y. and UEYANAGI, S. (1963) "Comparative distribution of eggs, larvae and adults in relation to biotic and abiotic environmental factors" *FAO Fish Rep.* **3(6)**, 979–1009.

YOSHIDA, H. O. and OTSU, T. (1963) "Synopsis of biological data on albacore *Thunnus germo* (Lacépède)" *FAO Fish Rep.* **2(6)**, 274–318.

ZEI, M. (1967) "The behaviour of *Sardinella aurita* Val. in relation to light and temperature" FAO Conf. Fish Behaviour, *FAO Fish Rep.* **62.2**, 469–475.

ZIJLSTRA, J. J. (1957) "The Dutch herring fisheries in 1955" *Ann. Biol. Cons. Perm. Intern. Explor. Mer.* XII, 200–204.

ZUPANOVIC, S. (1967) "Study of sardine (*Sardina pilchardus* Walb) behaviour in their natural environment by echo sounder and environmental factors" FAO Conf. Fish Behaviour, *FAO Fish Rep.* **62.2**, 269–282.

AUTHOR INDEX

SUBJECT INDEX

195